Kids and Credibility

This report was made possible by a grant from the John D. and Catherine T. MacArthur Foundation in connection with its grant making initiative on Digital Media and Learning. For more information on the initiative visit www.macfound.org.

The John D. and Catherine T. MacArthur Foundation Reports on Digital Media and Learning

Peer Participation and Software: What Mozilla Has to Teach Government by David R. Booth

The Future of Learning Institutions in a Digital Age by Cathy N. Davidson and David Theo Goldberg with the assistance of Zoë Marie Jones

The Future of Thinking: Learning Institutions in a Digital Age by Cathy N. Davidson and David Theo Goldberg with the assistance of Zoë Marie Jones

Kids and Credibility: An Empirical Examination of Youth, Digital Media Use, and Information Credibility by Andrew J. Flanagin and Miriam Metzger with Ethan Hartsell, Alex Markov, Ryan Medders, Rebekah Pure, and Elisia Choi

New Digital Media and Learning as an Emerging Area and "Worked Examples" as One Way Forward by James Paul Gee

Living and Learning with New Media: Summary of Findings from the Digital Youth Project by Mizuko Ito, Heather Horst, Matteo Bittanti, danah boyd, Becky Herr-Stephenson, Patricia G. Lange, C. J. Pascoe, and Laura Robinson with Sonja Baumer, Rachel Cody, Dilan Mahendran, Katynka Z. Martínez, Dan Perkel, Christo Sims, and Lisa Tripp

Young People, Ethics, and the New Digital Media: A Synthesis from the GoodPlay Project by Carrie James with Katie Davis, Andrea Flores, John M. Francis, Lindsay Pettingill, Margaret Rundle, and Howard Gardner

Confronting the Challenges of Participatory Culture: Media Education for the 21st Century by Henry Jenkins (P.I.) with Ravi Purushotma, Margaret Weigel, Katie Clinton, and Alice J. Robison

The Civic Potential of Video Games by Joseph Kahne, Ellen Middaugh, and Chris Evans

Kids and Credibility

An Empirical Examination of Youth, Digital Media Use, and Information Credibility

Andrew J. Flanagin and Miriam Metzger

with Ethan Hartsell, Alex Markov, Ryan Medders, Rebekah Pure, and Elisia Choi

The MIT Press
Cambridge, Massachusetts
London, England

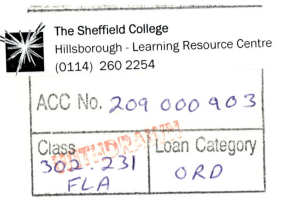
© 2010 Massachusetts Institute of Technology

For information about special quantity discounts, please email special_sales@mitpress.mit.edu.

This book was set in Stone Sans and Stone Serif by the MIT Press. Printed and bound in the United States of America.

Library of Congress Cataloging-in-Publication Data

Flanagin, Andrew J.
Kids and credibility : an empirical examination of youth, digital media use, and information credibility / Andrew J. Flanagin and Miriam Metzger; with Ethan Hartsell ... [et al.].
    p.   cm.—(The John D. and Catherine T. MacArthur Foundation reports on digital media and learning)
Includes bibliographical references.
ISBN 978-0-262-51475-0 (pbk. : alk. paper)
1. Mass media and youth—United States. 2. Digital media—United States—Social aspects. 3. Electronic information resources—United States.
4. Information behavior—United States. 5. Truthfulness and falsehood—United States. I. Metzger, Miriam J. II. Hartsell, Ethan. III. Title.
HQ799.2.M35F53   2010   302.23'10835—dc22   2009054316

10   9   8   7   6   5   4   3   2   1

Contents

Series Foreword

The John D. and Catherine T. MacArthur Foundation Reports on Digital Media and Learning, published by the MIT Press in collaboration with the Monterey Institute for Technology and Education (MITE), present findings from current research on how young people learn, play, socialize and participate in civic life. The Reports result from research projects funded by the MacArthur Foundation as part of its $50 million initiative in digital media and learning. They are published openly online (as well as in print) in order to support broad dissemination and to stimulate further research in the field.

Executive Summary

The enormous amount and variety of information currently available to people online present both tremendous opportunities and serious challenges. Readily available Web-based resources provide extraordinary promise for learning, social connection, and individual enrichment in a wide variety of forms. Yet, the availability of vast information resources also makes the origin of information, its quality, and its veracity less clear than ever before, resulting in an unparalleled burden on individuals to accurately assess information credibility.

Contemporary youth are a particularly intriguing and important group to consider with regard to credibility because of the tension between their technical and social immersion with digital media and their relatively limited development and life experience compared to adults. Although those who have grown up in an environment saturated with networked digital media technologies may be highly skilled in their use of media, they are also inhibited by their cognitive and emotional development, personal experiences, and familiarity with the media apparatus.

Despite these complex realities, examinations of youth and digital media to date have typically been somewhat simplistic. To provide a comprehensive look at children and online information credibility, this project employed a large-scale, Web-based survey of a representative sample of 2,747 children with Internet access in the United States, ages 11 to 18. In addition, one parent of each child was surveyed to obtain household indicators of digital media use, parental involvement, and various demographic factors.

Findings from this project constitute the first systematic survey of youth designed to assess their information-seeking strategies and beliefs across a wide variety of media and information types. As such, findings can be used to inform parents, educators, and policy makers interested in digital literacy and to understand the realities of children's relationship to digital media and the information they glean from such media.

Key findings of this project can be organized in terms of children's Internet usage, their perceptions of information credibility and factors affecting these perceptions, child/parent dyads and credibility assessments, and Web site exposure and evaluation.

Regarding children's Internet usage:

• The vast majority of children began using the Internet between second and sixth grades, with a majority of kids online by third grade. Nearly all kids (97 percent) are online by the eighth grade. Children use the Internet (not including email) for an average of almost 14 hours per week, and usage generally increases with age, from an average of 8 hours weekly among 11-year-olds to 16 hours per week for 18-year-olds.

• Overall, children rely fairly heavily on the Internet. The most important general uses include social networking, virtual usage (i.e., gaming and the like), information contribution in various forms (e.g., sharing files with others or creating personal Web sites, blogs, or journals), and commercial use (which is not very common among children). Although children generally acknowledge that information overabundance might pose a problem for them, nearly two-thirds of children report that their life would be either a little or much worse overall if they could not go online again, which is more pronounced with age.

• Children believe that they are highly skilled Internet users. Even 11-year-olds believe that their technical skill, search skill, and knowledge about Internet trends and features are higher than other Internet users.

• Seventy-five percent of parents control their child's access and use of the Internet by placing the computer in a certain location in the home, limiting the sites their child can visit, limiting the amount of time their child can go online, or controlling their children's Internet access in other ways. Parental oversight of children's online activities decreases as kids get older, with each method of control reported about half as frequently by parents of older children compared to parents of younger children.

Regarding children's perceptions of information credibility:

• Young people are concerned about credibility on the Internet, yet they find online information to be reasonably credible, with 89 percent reporting that "some" to "a lot" of information online is believable. While the amount of information they find credible increases with age somewhat, their concern about credibility does not.

• Their concern about credibility could stem from the fact that 73 percent of children have received some form of information literacy training, and the majority of parents report that they talk to their kids about whether to trust Internet information.

• A third of children reported that they, or someone they know, had a bad experience due to false information found on the Internet or through email. In addition, nearly two-thirds said that they had heard a news report about someone who had a bad experience because of false information online. These experiences affect how skeptical kids are of Internet information.

• Among several options, the Internet was rated as the most believable source of information for schoolwork, entertainment, and commercial information, as well as second most believable source for health information and third most believable for news information. Notably, children report that the Internet is a more credible source of information for school papers or projects than books.

• Kids are not very trusting of blogs, but they do find *Wikipedia* to be somewhat believable. Many children report believing information on *Wikipedia* substantially more than they think other people should believe it.

• Young people are appropriately skeptical of trusting strangers or people they meet online and are decidedly more trusting of people they meet in person.

• Children differentiate in reasonable ways among entertainment, health, news, commercial, and school-related information online when deciding which credibility assessment tools to use and with how much effort to employ them. Although this is generally encouraging, children also report finding entertainment and health information to be equally believable online, suggesting a suboptimal degree of skepticism between these diverse information types that have potentially quite different consequences.

• Older kids also show greater diversity and rigor in assessing the credibility of online information. Moreover, young people who are less analytic in their processing of information report trusting strangers online more and are more likely to be fooled by false information online.

• Children's concerns about credibility appear to be driven largely by analytic credibility evaluation processes, which involve the effortful and deliberate consideration of information. By contrast, actual beliefs about the credibility of information they find are dictated by more heuristic processes, by which decisions are made with less cognitive effort and scrutiny. This suggests that while most kids take the idea that they should be concerned about credibility seriously (by invoking a systematic and analytical approach), many also exhibit a less rigorous approach to actually evaluating the information they find online.

• There was no clear evidence of a "digital divide" in terms of the credibility beliefs and evaluations of kids from different demographic backgrounds. Instead, the rigor with which kids evaluate information they find online drives much of their credibility beliefs and concerns.

Regarding child/parent dyads and credibility assessments:

• Parents believe they are more adept at assessing credibility online than their children, and children almost universally share this assessment. This is particularly pronounced for younger children. However, the gap between parents and their children in this regard narrows with age.

• Children and adults both demonstrate an optimistic bias in their ability to identify credible information when compared to "typical" Internet users, indicating that they believe they are better equipped to discern information credibility than the average user. This is true even among children as young as 11 years old.

Regarding children's Web site exposure and evaluation:

• A majority of children displayed an appropriate level of skepticism when presented with hoax Web sites, a trend that contradicts prior research about this type of site. Nonetheless, approximately 10 percent of children still believed hoax sites either "a lot" or "a whole lot," indicating some lingering and important concerns.

• Children found encyclopedia entries that they believed originated from *Encyclopedia Britannica* to be significantly more believable than those they believed originated from either *Wikipedia* or *Citizendium*.

• The actual source of an online encyclopedia entry (i.e., taken from *Wikipedia*, *Citizendium*, or *Encyclopaedia Britannica*) was irrelevant to how credible the entry was found to be by children. However, encyclopedia entries were assessed as less believable when placed on *Wikipedia*'s site than when they were placed on the other sites. In addition, entries actually originating from *Wikipedia* were perceived as more believable when they appeared on *Citizendium*'s web page than if they appeared on *Wikipedia*'s page, and even more believable if they appeared to have originated from *Encyclopaedia Britannica*. Thus, ironically, while children find the content of *Wikipedia* to be most credible, they find the context of *Wikipedia* as an information resource to be relatively low in credibility.

▪ Children largely found product ratings to be credible and important to their assessments of commercial information. Average product ratings were significantly more influential than the number of ratings the product received, and there was some evidence that older children in particular were influenced slightly by the combination of average ratings and the number of ratings considered together.

Overall, this project provides a comprehensive investigation into youth's Internet use and their assessment of the credibility of online information. The findings—which are generalizable to households in the United States with Internet access—represent the current state of knowledge on this topic and serve as an important springboard for future research.

Acknowledgments

We are indebted to a great many people who contributed in various ways to this project.

First and foremost, we deeply appreciate the support of the John D. and Catherine T. MacArthur Foundation and in particular the vision, guidance, and intellect of Connie Yowell and Craig Wacker. The MacArthur Foundation's initiative on Digital Media and Learning (DML) has served as a remarkable resource for all those it has supported, and we have benefited immensely from the conversations and help of other DML participants.

We also want to thank Sandra Calvert and Paul Klaczynski, who served as consultants on this project.

A great deal of support has, of course, come from those at the University of California, Santa Barbara, as well. We are fortunate to have as our colleagues the doctoral students also listed on this report and, in addition, we have benefited enormously from the research assistance of a number of talented undergraduate students, including Jennifer Bryan, Jennifer Dossett, Westin Jacobsen, Kaitie Larsen, Cori Ochoa, Kamyab Sadaghiani, Caitie Ulle, and Arrington Walcott. Finally, Katie Bamburg, Jana

Bentley, and Monica Koegler-Blaha have provided invaluable support for this project through the Institute for Social, Behavioral, and Economic Research at UCSB.

Rationale and Overview

With the sudden explosion of digital media content and information access devices in the last generation, there is now more
information available to more people from more sources than
at any other time in human history. Pockets of limited access
by geography or status notwithstanding, people now have
ready access to almost inconceivably vast information reposito-
ries that are increasingly portable, accessible, and interactive in
both delivery and formation. One result of this contemporary
media landscape is that there exist incredible opportunities for
learning, social connection, and individual enhancement in a
wide variety of forms.

At the same time, however, the origin of information, its
quality, and its veracity are in many cases less clear than ever
before, resulting in an unparalleled burden on individuals to
find appropriate information and assess its meaning and relevance. Moreover, wide-scale access to information and the mul-
tiplicity of available sources also make it extremely complex to
assess the credibility of information accurately. And yet, it is
also highly consequential, since not having the skills to

accurately assess the credibility of information can have serious social, personal, educational, relational, health, and financial consequences in today's networked world.

While this is true for all users of digital media, youth are a particularly intriguing group to consider with regard to credibility because of the tension between their technical and social immersion with digital media and their relatively limited development and lived experience compared to adults. On the one hand, those who have literally grown up in an environment saturated with networked digital media technologies may be highly skilled in their use of media to access, consume, and generate information. This suggests that in light of their special relationship to digital tools, youth are especially well positioned to navigate the complex media environment successfully. Indeed, forms of credibility assessment that rely on information to be spread efficiently through social networks highlight some intriguing advantages for youth populations, who are often extremely interconnected compared to adults. In such instances, younger users may actually be better equipped than adults to transmit information pertaining to an entity's credibility quickly and efficiently via their social networks.

On the other hand, youth can be viewed as inhibited in terms of their cognitive and emotional development, life experiences, and familiarity with the media apparatus. This perspective suggests that although youth may be talented and comfortable users of technology, they may lack critical tools and abilities that enable them to seek and evaluate information effectively. Children's relative lack of life experience, for instance, may put them at greater risk than adults for falsely accepting a source's self-asserted credibility, since such assessments are based on accumulated personal experience, knowledge, reputation, and examination of competing resources. As a group, youth have fewer life experiences to which they might compare information than do most adults. In addition, youth may not have the same level of experience with or knowledge about media institutions, which might make it difficult for them to understand differences in editorial standards across various media channels and outlets compared to adults who grew up in a world with fewer channels and less media convergence. As a consequence, some youth may not have the same level of skepticism toward digital media or particular sources as adults do, because these media are not seen as "new" to younger users who cannot remember a time without them.

Although a good deal of scientific knowledge is accruing with regard to how people determine the credibility of information they get via digital media, extremely little of this work has focused on children. This is surprising, given the unique relationship of contemporary youth to media technology. We know, for example, that youth are more likely than adults to turn to digital media first when researching a topic for school or personal use; they are more likely to read news on the Internet than in a printed newspaper; and they are more likely to use online social networking tools to meet friends and to find information. In other words, the primary sources of information in their world are often digital, which is quite different from any prior generation.

Indeed, many have noted that their special relationship to digital media impacts the way youth approach learning and research. As the first generation to grow up with the Internet, young people are comfortable collaborating and sharing information via digital networks, and do so "in ways that allow them to act quickly and without top-down direction" (Rainie 2006, 7). Moreover, the interactivity afforded by networked digital media allows young people to play the roles of both information source and receiver simultaneously as they critique, alter, remix, and share content in an almost conversational manner using digital tools. These realities, we believe, have profound implications for how young people both construct and assess credibility online.

Despite these complex realities, examinations of youth and digital media have often been somewhat simplistic, focusing for example on the popular generation gap caricature, where youth are portrayed as either technologically adept compared to adults or as utterly vulnerable and defenseless. Such considerations fail to focus on the most important and enduring by-products of heavy reliance on digital media: the impact of "growing up digital" (Tapscott 1997) is that more and more of the information that drives our daily lives is provided, assembled, filtered, and presented by sources that are largely unknown to us, or known to us in nontraditional ways. Yet, we have only begun to explore what this means for younger users who are not only immersed in digital media now but will be for the entirety of their lives.

To address these issues, this project provides a comprehensive look at kids and online information credibility, using a large-scale survey of children in the United States, ages 11 to 18. The research reported here fills the current void in knowledge about how youth seek information and assess credibility using

many types of digital media. In the face of increasing disinter-
mediation and media complexity, the practical application of
such knowledge could be employed to empower users to reap
the benefits of the vast digital information environment while
minimizing the risks of relying on information that may be mis-
leading, incomplete, or wholly inaccurate. Overall, data from
this survey constitute the first systematic study of youth
designed to assess their information-seeking strategies and
beliefs across a wide variety of media and information types. As
such, our findings offer unprecedented insight into how young
people think about credibility today.

Findings from this study can be used to inform parents, educators, and policy makers interested in digital literacy, and to
understand the complex realities of children's relationship to
digital media and the information they glean from them.

Research Approach

Overview

Although there is a burgeoning literature and empirical work on adults and credibility (see Flanagin and Metzger 2007), including informative work on college-age adults (Metzger, Flanagin, and Zwarun 2003; Rieh and Hilligoss 2007), extremely little research has been conducted on pre-college age youth. What empirical research does exist is almost exclusively based on interviews of very small samples of children and adolescents, which cannot be generalized with any accuracy to the overall youth population. To redress this shortcoming, this project generated survey data from a representative sample of young people in the United States.

The survey instrument used in this study was generated through a multi-step, multi-method process. The initial survey topics were based on an extensive review of past literature and existing surveys on information trust, credibility, and quality. To better understand cognitive and developmental issues relevant specifically to youth information assessment and processing, research experts currently working in the fields of developmental psychology and cognitive psychology were recruited as project consultants. A draft version of the questionnaire was critiqued and modified through working sessions with these consultants over the course of multiple days. The outcome of these sessions was a comprehensive questionnaire pertaining to digital media use and assessment, informed by contemporary perspectives from cognitive and developmental psychology.

To gauge the clarity, comprehensiveness, and relevance of the questionnaire for youth audiences, a small-scale focus group was next conducted among children ages 9 to 18. Questionnaire modifications were again made based on this session, and the consultants' recommendation that participants should be no younger than 11 years old was confirmed. Next, to further validate the questionnaire for youth audiences, as well as for the portion of the survey evaluating parents' assessments of their child's online information behaviors, 40 parent-child pairs were recruited. These pairs represented a broad range of races, ethnicities, and household incomes, as well as roughly equal numbers of children in each age cohort and sex. Members of each child/parent dyad underwent a separate hour-long face-to-face interview with researchers, in which they provided feedback on questionnaire content, question wording, and general survey administration. Once again, this feedback was used to modify the questionnaire.

This version of the survey was then pilot-tested among 183 undergraduate college students, in order to gauge the reliabilities of the attitudinal and usage scales included in the questionnaire and to uncover any other outstanding issues. Minor adjustments were made to the questionnaire, which was then forwarded to the research firm that administered the survey to the target population, as noted in more detail below.

Given the near-ubiquitous use of the Internet among contemporary youth, and the fact that this constituted our target audience, Web-based survey techniques were used to assess youths' Web usage behaviors and attitudes about online credibility. The questionnaire was administered to a sample of youth with Internet access in the United States, ranging in age from 11 to 18 years old. In addition, one parent of each child was surveyed to obtain household indicators of digital media use, parental involvement in their child's digital media use, and demographic factors.

The 2,747 valid responses obtained were a roughly equal representation across youth age cohorts (i.e., approximately 340 respondents for each age within the range). Surveying a range of ages accomplished a number of things: it represented children at critical junctures in social and cognitive development; it considered youth at times in their academic development and in their development as citizens that are key to their future decisions and choices; and it enabled comparisons between children of various ages, providing relatively precise comparisons across age cohorts (e.g., junior high versus high school), to pinpoint the key junctures at which children attend to, and act on, distinctions in information credibility. Moreover, this sample size yielded sufficient representation across sex and other demographic differences to facilitate comparison across these factors. Finally, because the current project is an extension of ongoing research on adults, it will also provide direct comparison between youth and adult populations in future studies, which will suggest lifespan differences in the key variables of interest.

Survey Methodology

Survey Administration

The survey was conducted online by the research firm Knowledge Networks and was fielded between June 17 and July 26, 2009. Knowledge Networks maintains a probability-based panel of participants and is thus the only online survey source that meets the standard of federal and peer review, setting the gold standard in the industry. As mentioned earlier, 2,747 children in the United States between the ages of 11 and 18 who use the Internet and who live at home, as well as one parent for each child participant, completed the survey. Statistical results were weighted to correct known demographic discrepancies between the U.S. population and Knowledge Networks' online panel. Details on the design, execution, and weighting procedures of the survey are discussed below. Additional information about the survey methodology and subject panel used by Knowledge Networks can be found in Appendix B.

Sample Design

Knowledge Networks has recruited the first online research panel that is representative of the entire U.S. population. Panel members are randomly recruited by probability-based sampling (telephone, mail-, and Web-based surveys), and households are provided with access to the Internet and hardware if needed (although this did not apply to the current survey, since our target sample included only current Internet users). After initially accepting the invitation by Knowledge Networks to join the panel, respondents are then profiled online by answering demographic questions, and maintained on the panel using the same procedures established for research subjects recruited by random digit dialing. The sample for this study was drawn from a combination of random digit dialing and address-based sampling methods (taken from the U.S. Postal Service's Delivery Sequence File). The combination of these two frames allows Knowledge Networks to reach homes without a landline telephone, homes with numbers on the do-not-call list, and homes that use call-screening that normally would be missed by random digit dialing methods alone.

The typical survey commitment for Knowledge Networks panel members is one survey per week or four per month, with a duration of 10 to 15 minutes per survey. Knowledge Networks' general sampling rule is to assign no more than one survey per week to members. Knowledge Networks operates an ongoing, modest incentive program to encourage participation and create member loyalty. Members can enter special raffles or can be entered into special sweepstakes to win both cash and other prizes.

For this study, households with children living at home between 11 and 18 years of age were identified by Knowledge Networks within their online panel (18-year-olds not living at home were excluded from this sample). A sample was drawn at random from among active panel members. For this survey, 5,936 U.S. adult parents with at least one child age 11 to 18 were selected for the main and pretest surveys.

Contact Procedures

Potential participants received a notification email letting them know there is a new survey available for them to take. This email notification contained a link that sent them to the survey questionnaire. No login name or password was required.

Parents were first asked to complete a short screening questionnaire to confirm that they had a child age 11 to 18 and to gain consent for the child to participate. Upon completion of their portion of the survey, parents were asked to have one selected 11- to 18-year-old complete a longer series of questions designed to assess the child's use of the Internet. To accommodate participants' schedules and increase the chances of having a child complete the survey, parents were told that they could have their child complete the survey at a later time if that was more convenient.

A first email reminder was sent to all non-responding panel members in the sample on July 2, 2009. Second and third email reminders were sent 7 and 12 days later, respectively. Finally, calls were made to all remaining non-responding panel members starting July 16, 2009 and throughout that weekend.

Incident and Completion Rates

For this survey, 3,136 adult parents with at least one child aged 11 to 18 responded to the invitations, representing a 52.8 percent completion rate. 2,747 parent-child (aged 11 to 18) pairs completed the survey and qualified for analysis, representing a 91.7 percent qualified rate or 46.3 percent response rate.

Sample Weighting

The survey responses were weighted to provide results that are generalizable to the U.S. population of Internet households. Two weighting strategies were employed to compensate for non-response and other sources of survey error that might bias the results.

First, a post-stratification adjustment using demographic distributions from the most recent U.S. Census Bureau's Current Population Survey data was used to balance errors due to panel recruitment methods and panel attrition. Demographic variables used for this weighting included gender, age, race, education, and Internet access.[1] This weighting was applied before the selection of the sample was made for this study.

In addition, a study-specific post-stratification weight was applied after data collection to adjust for the study's sample design and survey non-response. A weight was calculated for all qualified children to make them comparable to 13- to 18-year-olds who have Internet access at home.[2] Household income was also included as a weighting variable since education could not be included (i.e., most of the children in this age range have less than a high school education). The sample design effect for this weight is 1.58.

Sample Characteristics

This section provides a detailed profile of the demographic characteristics for both the parent and child samples.

Parent and Household Demographics

Parents in the sample were 45 years old on average (standard deviation = 7.25). Most had attended college, with 53 percent having had at least some college, and 47 percent earning a bachelor's degree or higher. In terms of race, 75 percent of the parents were white; 9 percent were black, non-Hispanic; 9 percent were Hispanic; 4 percent were other, non-Hispanic; and 4 percent reported their race as Mixed, non-Hispanic. Thirty-one percent of the participating parents were male, and 69 percent were female. Eighty percent of parents were married or living with a partner, 20 percent were divorced, separated, widowed, or never married. Seventy-six percent of parents were working at the time that the data were collected for this study.

Household annual income ranged from less than $5,000 to more than $175,000, with an average income ranging from $60,000 to $85,000. Most families (88 percent) had between 3 and 5 members living in the household, and the average number of children living at home was 2.25 (standard deviation = 1.39). Participants came from all parts of the United States, with slightly more coming from the Midwest (31 percent) compared to the Northeast (19 percent), South (28 percent), and West (23 percent). Table 1 shows the more specific breakdown of the sample's geographic distribution:

Child Demographics

The child respondents consisted of 53 percent males and 47 percent females ranging in age from 11 to 18 years, with an average age of 14.33 (standard deviation = 2.28). Table 2 shows the percentage of children in the sample within each age group surveyed. Seventy-five percent of the child respondents reported that they were white; 9 percent were black, non-Hispanic; 12 percent were Hispanic; 0.4 percent were other, non-Hispanic; and 4 percent reported being Mixed race, non-Hispanic.

Table 1

Percent of participants from various U.S. geographic areas

Region	Percent Residing
New England	4.9
East–North Central	20.6
East–South Central	4.7
Mid-Atlantic	13.6
South Atlantic	15.1
Mountain	7.2
Pacific	15.5
West–North Central	10.7
West–South Central	7.7

Table 2

Number and percent of participants within each age category sampled

Age	Number	Percent
11	378	13.8
12	371	13.5
13	385	14
14	323	11.8
15	327	11.9
16	316	11.5
17	368	13.4
18	279	10.2

Research Findings

Internet Usage among Youth

To better understand how young people may be affected by the Internet, we first assessed the prevalence and nature of their Internet usage. We examined children's general use of the Internet by measuring when they first started using the Internet and how often they go online. We also investigated parental control of their child's access to and use of the Internet, since children's use behaviors are in some cases not entirely under their own control. We next asked a number of questions about children's online activities, in order to fully understand their usage behaviors and activities. Finally, we evaluated young people's perceptions of their Internet skill level and their overall impressions of the Web environment.

Children's Use of the Internet
The vast majority of children began using the Internet at some point between second and sixth grade, with a majority of kids (52 percent) online by third grade. Nearly all kids (97 percent) reported being online by the time they were in eighth grade.

Children in the study reported that they use the Internet (not including email), for an average of 13.53 hours per week, although there was quite a lot of variation in the amount of time they spent online (standard deviation = 12.44). This means that the majority of kids (nearly 70 percent) spent anywhere from 1 to 26 hours a week online, with few spending less than one hour or more than 26 hours online weekly.

Internet usage did, however, generally increase with age, with 11-year-olds reporting an average of 8.21 hours (standard deviation = 6.06) per week online, which doubled by age 18 to 16.38 hours per week on average (standard deviation = 11.63). Figure 1 shows the average amount of time per week spent online by children of different ages.

Parental Control of Children's Access to and Use of the Internet

A large majority of parents (75 percent) control their child's access and use of the Internet in some manner. In our total sample, most parents (53 percent) place the computer in a certain location in the home in order to keep an eye on what their child is doing online. Forty-three percent limit the sites their child can visit, 42 percent limit the amount of time their child can go online, and 19 percent control their children's Internet access in other ways. Among only those parents who control their child's access to the Internet in some fashion (the 75 percent of the sample mentioned earlier), 71 percent place the computer in a certain location in the home, 57 percent limit the sites their child can visit, 55 percent limit the amount of time their child can go online, and 25 percent control their children's Internet access in other ways.

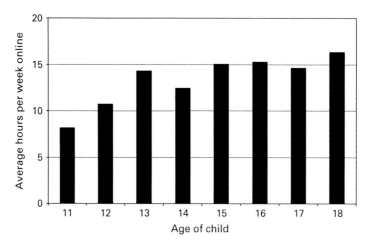

Figure 1
Weekly usage of the Internet by age

However, parental oversight of children's online activities decreases as kids get older. For example, 94 percent of parents of 11-year-olds report that they control their child's Internet access and use, whereas only 45 percent of parents of 18-year-olds report controlling their child's online activities. In fact, nearly each type of control (e.g., placing the computer in a certain location in the home, limiting the sites their child can visit, and limiting the amount of time their child may go online) is used about half as frequently by parents of older children compared to parents of younger children, as seen in figure 2.

In addition, about two-thirds of parents reported that they sit with their children while they go online, with only 38 percent reporting doing this "never" or "rarely." Nearly half (47 percent)

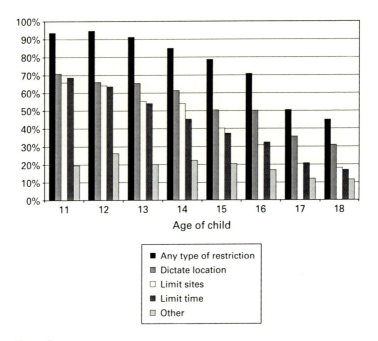

Figure 2
Percentage of parents restricting their child's Internet access by age of child

reported sitting with their child "sometimes," whereas 16 percent reported sitting with their child "often" or "very often." As with the other forms of restrictions we examined, the percentage of parents who sit with their children while they are online decreases with age, with a majority of parents of 11-year-olds sitting with their children "often" or "sometimes," while a majority of parents of 17- to 18-year-olds report doing this "rarely" or "never."

Children's Online Activities

To assess what young people are doing online, we asked a series of questions about how often they use the Web for a variety of purposes. Overall, children reported using the Web most often for watching videos and interacting with others through social networks, noting that on average they do each of these activities between "sometimes" and "often" (3 and 4 on a 5-point scale, respectively). They also reported that they look up information on *Wikipedia*, play games and use avatars, and buy things online between "rarely" and "sometimes." Other activities, including posting information to various groups, sharing video or music files, and bidding in online auctions, were also reported, but were reported only "rarely" or less often on average. Figure 3 shows children's usage of Web-based information resources, arranged in order of highest average reported uses to the least commonly reported uses.

Although the frequency of most uses of the Web was relatively consistent across ages, some notable trends were observed in children's Web usage by age. Many uses of the Web increased with age, or increased up to a certain age, including sharing videos and music, posting original artwork, photos, stories, or videos online, and using social networking sites, as represented in figures 4, 5, and 6. In each of these cases, there appears to be a notable plateau in the frequency of these uses that occurs at or about the age of 15. Interestingly, there was actually a *decrease* in the extent to which children reported visiting virtual worlds like *Second Life* or *World of Warcraft* as they grew older, as shown in figure 7.

Some of these usage trends can be better understood in terms of more general usage behaviors that emerge from the specific

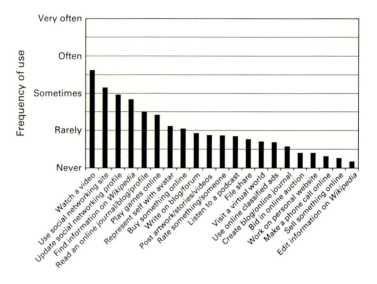

Figure 3
Children's use of Web-based information resources

Web uses reported by respondents. To assess this, we constructed several scales that describe children's types of Web usage.[3] We derived measures describing (a) the frequency of children's social network use, (b) the extent to which they contribute information to others, (c) the degree to which their use of the Web is virtual (for activities like playing games), and (d) the degree to which they use the Web for commercial purposes.

Social network use was based on how often children use social networking sites, read others' blogs, profiles, or journals online, and update their own social networking profiles (Cronbach's alpha = 0.87). Subsequent analyses showed that older children were more likely to be heavier users of social networks, as were

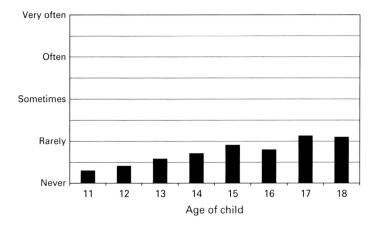

Figure 4

Frequency of file sharing by age

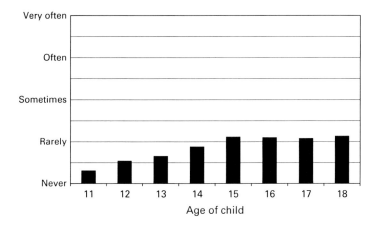

Figure 5

Frequency of posting original content by age

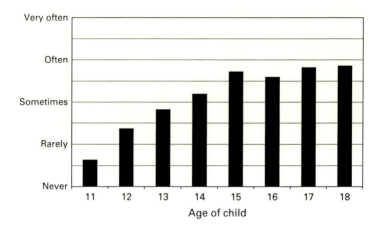

Figure 6

Frequency of using social networking sites by age

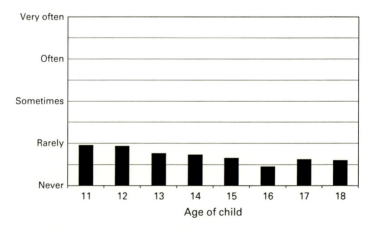

Figure 7

Frequency of visiting virtual worlds by age

more skilled users, girls, and those who spent more time online per week. Children who earned lower grades in school and had less parental control of their Internet access and use also showed higher use of social networking sites. In addition, those relying on group processes to discern credibility and those with "heuristic" decision making styles (as opposed to more analytical styles, covered in a later section of this report, "Factors Affecting Children's Credibility Evaluations") were also heavier social network users.

Information contribution scores were derived from such things as the extent to which children reported creating personal Web sites, blogs, or journals, posting information to groups or sharing files with others, or rating people or things online, through sites like Amazon, eBay, IMDb, or YouTube (Cronbach's alpha = 0.77). Greater Internet skill, reliance on group processes for making credibility decisions, time online per week, and age resulted in greater information contribution. In addition, among other factors, grades in school were negatively related to information contribution, girls were more likely to contribute, and high Internet social trust led to greater contribution, although higher general social trust led to lower information contribution.

Virtual uses consisted of representing oneself with an avatar, visiting virtual worlds, or playing games with others who are also online (Cronbach's alpha = 0.71). The strongest predictors of heavy virtual uses were hours online per week, sex (with girls more likely to engage in virtual uses), Internet skill, and age. In addition, virtual users tended to rely on group processes more in making credibility assessments, had been taught at some point about credibility issues online, tended to get good grades, and had been online a long time.

Finally, *commercial use* was composed of bidding in online auctions, buying or selling merchandise online, and frequenting sites like Craigslist.org to look at classified ads (Cronbach's alpha = 0.69). Those with higher Internet skill and who were older use the Web for commercial purposes more than do kids with lower skill and who are younger, and commercial use of the Internet decreased as parents controlled their children's Internet access and usage more. Boys tended to use the Web for commercial purposes more than girls, and commercial users tended to rely on others and also invoke analytic methods when determining credibility online, though they also had high faith in their own intuition in determining what, and who, is trustworthy.

The frequency of social networking behaviors increases rather dramatically between the ages of 11 and 15, as shown in figure 8, after which it remains relatively constant. Whereas 11-year-olds report using social networking sites only between "never" and "rarely" on average, older teenagers use them on average between "sometimes" and "often." This confirms findings from past studies that have shown increases in social network site usage between younger (12- to 14-year-olds) and older (15- to 17-year-olds) kids (Lenhart and Madden 2007), and indicates that as children grow older they may be using more digital forms of communication to connect and interact within their social networks.

Largely consistent with our earlier observation about visiting virtual worlds, figure 9 shows that the frequency with which

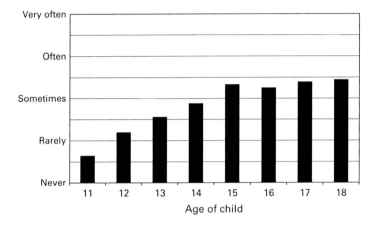

Figure 8
Frequency of social networking use by age

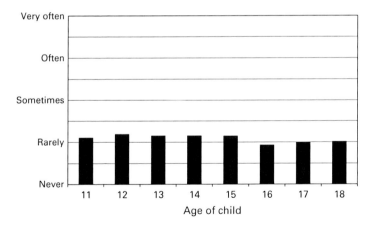

Figure 9
Frequency of "virtual" Web use by age

children's use of the Web for "virtual" purposes varies little by age. Contrary to popular conceptions of older teens as more frequent participants in virtual worlds, our data suggest that children's use of the virtual features of the Web is not only fairly low (indicating that they "rarely" use the Web in this way), but that this low usage is consistent across age. Not surprisingly, commercial use of the Web increases with age, as shown in figure 10, likely due to the availability of greater discretionary income and increased independence from parents.

Figure 11 shows differences in how often children use the Web for contributing information in various forms to others. Once again, these results suggest that as children get older they contribute more information, although it should be noted that information contribution remains rare overall, with average contributions of even older children not quite reaching even a

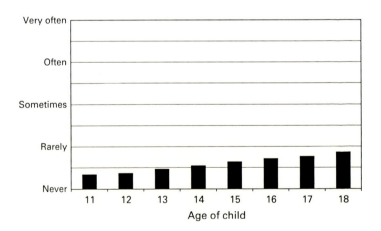

Figure 10
Frequency of commercial Web use by age

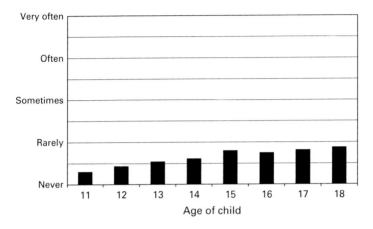

Figure 11

Frequency of online information contribution by age

level of doing so "rarely." Nonetheless, sharing information in various ways (via personal Web sites, blogs, or journals; posting information to groups or other people, or rating people or things online) is indicative of one of the most notable features of the Internet—its ability to enable information consumers to simultaneously be information providers, a behavior that appears to increase slightly during childhood.

Together, these findings show that children are going online for a variety of purposes, although to varying degrees according to their age. To evaluate the extent to which they have become reliant on the Internet, we asked them to assess how their lives would be different if they could never use the Internet again. Only 1 percent reported that their lives would be better overall if they could not go online again. By contrast, 64 percent reported that their life would be either much or a little worse

overall, and 23 percent said that it would not make any difference to their life overall if they could not go online again. In
general, as their age increases children report that their lives
would be progressively worse without Internet access, indicating that older children experience increased reliance on the
Internet.

Children's Web and Internet Skill Levels

To evaluate children's self-perceptions of their skill levels with
Internet technologies, we asked them three questions, designed
to assess their technical skills (e.g., fixing connection problems
or changing computer settings), search skills (i.e., ability to find
what they are looking for online), and knowledge about Internet trends and features. All questions were rated on a 0 to 10
scale, where 0 indicated that they were much less skilled/knowl-
edgeable than other Internet users and 10 indicated that they
were much more skilled/knowledgeable than other Internet
users. The scale midpoint of 5 indicated that respondents
thought they were "about as knowledgeable/skilled as other
Internet users."

Results show that, overall, average perceived skill levels were
high. For technical skill, search skill, and knowledge about
trends and features, even 11-year-olds perceived that they were
on average more skillful/knowledgeable than other Internet
users (average scores = 5.32, 5.36, and 6.85, respectively). Moreover, across all three measures of Internet skill, children of all
ages perceived themselves to be more skillful than other Internet users (average scores were all above the scale midpoint), and
saw their search skills as significantly better (average = 7.47)

than their technical skills (average = 6.49) or knowledge of Internet trends and features (average = 6.48).

In addition, a trend emerged for all skill measures where between ages 11 and 15 children felt they were generally more skillful on average with each passing year. However, for all three skill measures, self-perceived average skill and knowledge typically *decreased* slightly after the peak at age 15, although generally not to levels that achieved statistical significance. Overall, this shows that children's average self-perceived skills appear to rise until the age of 15 or so, after which they generally either level off or decrease modestly.

There were also intriguing sex differences with regard to skill self-assessments. Setting aside children's ages for the moment, boys rated themselves as significantly more skillful on average than did girls, for both search skills and technical skills (but not for knowledge of Internet trends and features). Considering the age of respondents, however, paints a slightly more nuanced picture of children's skill assessments: 11-year-old boys rated themselves on average as more technically skillful than 11-year-old girls did; 12- and 14-year-old boys and girls did not differ on any self-assessed skill measure; and boys aged 15 through 18 tended to rate themselves as more skillful than did 15- to18-year-old girls, on average, on at least one measure of skill, and sometimes on all three measures. Thirteen-year-old girls, however, are distinct in that they rated themselves as significantly more skillful than 13-year-old boys rated themselves, on both search skills and on knowledge of Internet trends and features. Overall, meaningful sex differences on self-perceived skill show up as children reach their mid-teens, when boys believe themselves to be more highly skilled than girls do. Prior to that time, differences in skill are negligible other than a surprising peak among 13-year-old girls in terms of their perceived Internet and Web skills.

Children's Impressions of the Web Environment

When asked about the amount of information available online, children generally acknowledged that information overabundance might pose a problem for them. Specifically, children were asked, on a scale of 0 (not enough information online) to 10 (too much information online), what they thought about the amount of information available online overall. The average response was 7.57, indicating that they thought there was generally too much information available online. Moreover, 61 percent of children indicated responses ranging from 6 to 10 (too much information), whereas only 13 percent gave responses ranging from 0 to 4 (not enough information). 26 percent of children, however, did indicate that the amount of information online was "just right." There were no statistically significant differences in opinions about the amount of information available online by age or sex, indicating that this finding applied equally to children of all ages and both sexes.

Summary

Overall, findings indicate that children are actively using the Internet for a variety of reasons. Other than the reported frequency with which children's use of the Web is virtual, it seems that use of the Internet for myriad purposes increases with age and, therefore, with the total amount of time spent online.

This increase in time spent online with age may be due to the corresponding reduction of parental control. Interestingly, although many children said there is too much information online, the majority of children also indicated their Internet skill was fairly high, suggesting that while children may be overwhelmed or bombarded with information, they are confident in their ability to decipher and filter through it. This suggests that children believe they are relatively capable of successfully navigating the digital world, a premise that we explore in much greater depth in the remainder of this monograph, particularly as it relates to information credibility.

Perceived Trust and Credibility of Web-Based Information

This section and the next discuss survey results intended to assess children's perceptions of the credibility of information on the Web. We begin by considering general issues regarding the credibility of Web-based information, including children's trust of people they meet online and their experiences with false information on the Internet. We then look at differences in children's perceptions of online credibility across various types of information and information sources, as well as across several media (e.g., Internet, television, newspapers), and we investigate the behaviors children engage in when deciding which information and what sources to trust.

General Issues Regarding the Credibility of Web-Based Information
In general, children found information on the Web to be relatively believable, with 59 percent reporting that "some"

information was believable, and 30 percent reporting that "a lot" of the information found online was believable (see figure 12). There was also a small but significant tendency for perceived information believability to increase with age: 18-year-olds found more of the information online to be credible than 11- through 14-year-olds.

When asked how often they think about credibility, as well as how concerned others should be about the credibility of online information, children showed a healthy level of concern about these issues. Seventy-nine percent of children in the sample said they think about whether they should believe information they find online "sometimes" or more often, and 71 percent said that people should be "somewhat" to "very" concerned about the believability of online information. Figures 13 and 14 summarize these results. Age did not matter much in these findings, although 18-year-olds felt people should be more

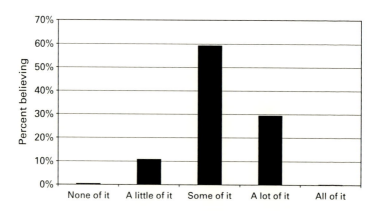

Figure 12

Amount of online information that children find credible

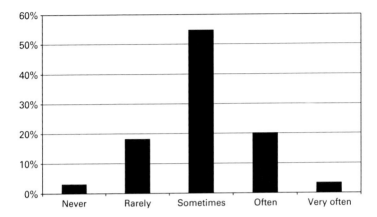

Figure 13
Frequency of thinking about credibility when online

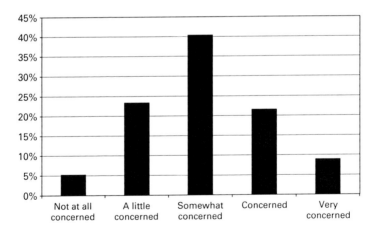

Figure 14
How concerned others should be about credibility

concerned about how believable information online is than both 12- and 14-year-olds.

Trust and the Negative Consequences of False Information Online
When asked whether they felt people could be trusted, children reported that they trusted people they knew or met in person more than they trusted people online. Specifically, on a 4-point scale (where higher scores indicate higher trust), children reported an average trust score of 2.65 for people in person, whereas online this score decreased to 1.17. Figures 15 and 16 show the specific differences in scores, as well as the categories of children's responses to the trust questions.

Similar to this, children tended not to trust strangers they meet online (e.g., in social networking sites, forums, online

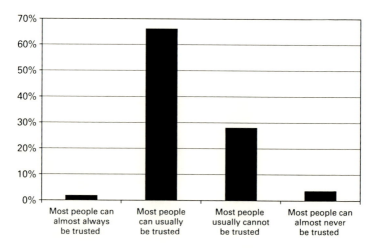

Figure 15
General social trust

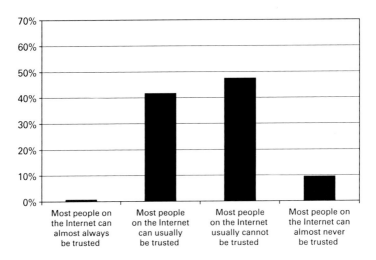

Figure 16
General Internet social trust

communities, etc.), with nearly half indicating they do not trust people under these circumstances at all, and only about 2 percent saying they trust strangers online "a lot" or "a whole lot." About a third of children did indicate they trust strangers online "a little bit" and 15 percent said they trust them "some" (see figure 17). It should be noted, however, that nearly 42 percent of the children in the sample indicated that they have "never" met a stranger online. There was little variance in children's level of stranger trust by age, although older children (16–18 years old) indicated they were slightly more trusting of strangers online than younger children (11–14 years old). Figure 18 illustrates these findings.

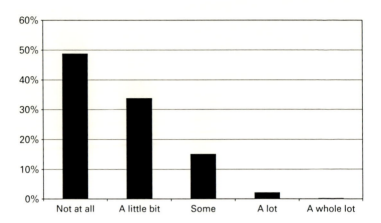

Figure 17
Degree that children trust strangers they meet online

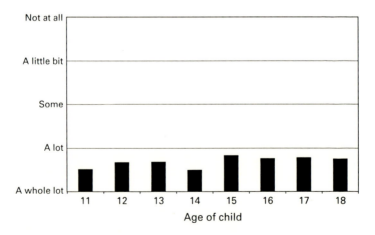

Figure 18
Degree that children trust strangers they meet online by age

One possible reason for these diminished levels of trust online is negative experiences in children's pasts. To assess this, we explored the extent to which children had negative experiences online, either firsthand or through others' experiences. Thirty-two percent of children reported that they, or someone they know, had a bad experience due to false information found on the Internet or through email. This did not vary by age. Nearly twice as many children (62 percent) reported that they had heard a news report about someone who had a bad experience because of false information online. In this case, differences were only found between 11- and 18-year-olds (with 18-year-olds reporting higher scores).

To mitigate or avoid negative experiences, children can of course be *instructed* in recognizing bad information or evaluating information in general. To assess the extent to which children have been taught about various issues regarding the credibility of online information, we asked two questions. First, we asked children if they had ever had someone (like a teacher, parent, librarian, or friend) teach them ways to decide what information from the Internet they should believe. Results show that 73 percent have indeed been instructed by someone on how to assess the credibility of information online.

Second, we asked parents how often they talk with their child about whether information on the Internet is trustworthy. Most parents (84 percent) reported that they talk with their children about whether the information on the Internet is trustworthy at least occasionally, with only 16 percent reporting doing this "never" or "rarely." Forty-five percent reported talking with their child "sometimes," and 31 percent reported

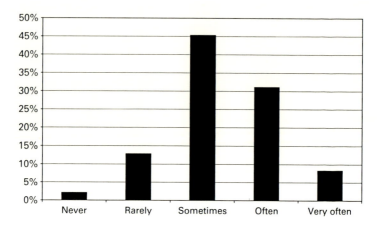

Figure 19

Frequency that parents talk with their children about online credibility

talking with their child "often." However, only 8 percent talk with their child "very often" (see figure 19). Interestingly, this does not change much with the age of the child, as a majority of parents within each age category say they talk with their children "sometimes" to "often" about whether information on the Internet is trustworthy.

Credibility Differences by Information Type and Source of Information

Past research on credibility has found that the degree to which adults believe information they find online varies by the type or topic of information they might search for. For example, people tend to be less likely to find commercial information or information coming from special interest groups to be credible, presumably because they recognize the strong potential for bias

(Flanagin and Metzger 2000, 2007). We wanted to see if similar patterns were found for younger Internet users. In addition, because most past work has focused on the credibility of Web sites, we wanted to explore young people's perceptions of newer information sources that they likely encounter online, including blogs and *Wikipedia*.

Information Type   Despite indications that children tend not to be terribly trusting of others online (as noted earlier), they indicate a great deal of faith in the Internet as a source of consequential information, compared to more trivial information pursuits. We asked children how likely they are to believe information on the Internet about a number of topics or types of information, including health or medical issues, news, something they may want to buy, entertainment information (e.g., about movies, musicians, celebrities, etc.), other people they meet online, and information they find for school papers or projects. Results showed that children varied in their likelihood of believing information across these topics.

Specifically, children were on average most likely to believe information on the Internet about schoolwork, followed by news, then entertainment and health information (which children were on average equally likely to believe), commercial information, and information about people they met online. Figure 20 illustrates these results. Although there were minor age differences with these findings (e.g., some of the older children did not distinguish between the believability of health and commercial information, and children of some ages did not distinguish between commercial and entertainment information), the general pattern of findings endured regardless of age.

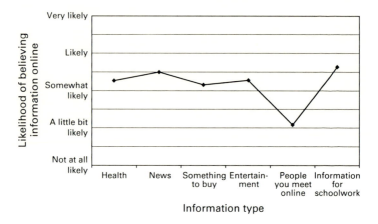

Figure 20
Credibility of online information by information type

Although it makes sense that news information, for example, is highly regarded in terms of its credibility, it is potentially problematic that children believe health and entertainment information equally. Indeed, these findings suggest that kids may not be processing the credibility of health information with any more rigor than they are assessing the trustworthiness of entertainment information, in spite of the fact that the possible negative consequences of believing false health information are far greater in scope and scale than are the negative consequences of false entertainment information. Nonetheless, there is also encouraging news, inasmuch as children tend to rate as least believable information that might be presented with persuasive, or even potentially nefarious, intent (i.e., commercial information and information about people they meet online).

Yet, it is unclear from these data whether children view information found for school as highly credible because of a high level of diligence in searching for or verifying it, or if they base such assessments on the convenience of finding information on the Internet.

Past research on college-age students (Metzger, Flanagin, and Zwarun 2003), however, shows that students not only rely on Web-based information for their schoolwork quite heavily, but verify its veracity less than adults, suggesting that convenience may be a more critical factor here.

Blogs Unlike their parents, children today are growing up in a news media environment full of sources that do not have the journalistic checks present in traditional media like newspapers, magazines, or radio and television broadcasts. Our survey assessed children's level of trust in both old and new information sources, with some interesting results.

Overall, kids do not find (news) blogs to be very credible. Seventy-nine percent say they are either "much less" or "somewhat less" believable than newspaper and television news. This does not vary much by age. It should be noted, however, that many kids were unsure about the comparative credibility of blogs and mainstream news, with 37 percent of all kids answering "I don't know" about their relative credibility and 8 percent of the total sample indicating that they did not know what a blog is. Figure 21 illustrates responses for those with an opinion on the topic.

Wikipedia Another recent information source phenomenon is *Wikipedia*, the online encyclopedia to which anyone can con-

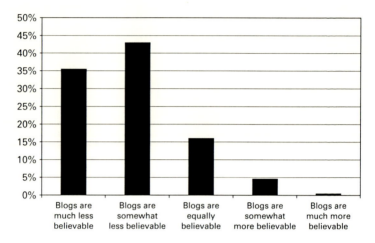

Figure 21
Credibility of blogs compared to newspapers and television

tribute information anonymously. *Wikipedia* currently boasts
more than 3 million entries and is among the top 10 most popular Web sites in the United States (Alexa 2009; Quantcast 2009).

Nearly all kids (99 percent) who completed our survey had
heard of *Wikipedia*, and the vast majority of them (84 percent)
have used it to look up information, with most reporting they
"sometimes" look up information on *Wikipedia*. However, few
(12 percent) have ever written or changed some information in
Wikipedia and those who have done so report doing it only
"rarely." This does vary by age, to some degree: older kids are
about 10 percent more likely to have done both activities (look
up information and write or change information in *Wikipedia*)
than the youngest kids in our sample.

However, when asked to identify what *Wikipedia* is from a list of seven possibilities (e.g., whether it is an online encyclopedia where anyone can contribute information, a social networking site, a Web site where you can play games, an online encyclopedia where only experts may contribute information, a company that sells books online), 9 percent admitted that they do not know what it is, and only 78 percent made the correct identification. Moreover, there was a small tendency for older kids (ages 16+) to more accurately understand what *Wikipedia* is. Because this distinction is important, in subsequent analyses we assessed only those children who correctly identified what *Wikipedia* is and how it operates.

Overall, children find *Wikipedia* to be fairly believable. Most children believe information from *Wikipedia* at least "some" (43 percent) or "a lot" (28 percent). However, children were slightly more skeptical about how much people should believe *Wikipedia*, with 23 percent saying it should be believed "a little bit," 49 percent saying it should be believed "some," and 20 percent saying it should be believed "a lot." Indeed, the extent to which children say people should believe information in *Wikipedia* is significantly lower than they report believing it themselves. These results are illustrated in figure 22. There were no differences between younger and older children either in how much they themselves believed or in how much they thought that people should believe the information found on *Wikipedia*.

Differences in Credibility across Media

To determine which channel of information delivery children think provides the most believable information for a variety of

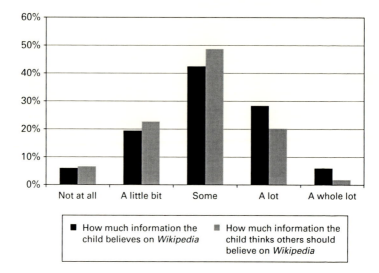

Figure 22
Credibility of *Wikipedia*

purposes, we asked them to indicate which among several alternatives (including the Internet, television, books, magazines, newspapers, radio, and someone they talk to in person) provides the most believable information.

Consistent with past research (Flanagin and Metzger 2000), differences emerged across technologies depending on the type of information sought. When looking for health or medical information, 39 percent of children indicated that they would most believe someone they talk to in person, followed by the Internet (21 percent) and books (20 percent), which were roughly equivalent. Children indicated that the most believable news information originated from television (54 percent), followed by newspapers (24 percent), and then the Internet (11 percent). Commercial information was best retrieved from the Internet (41 percent) or in person (33 percent), followed by television and magazines (10 percent each). The most believable entertainment information, according to children, can be found on the Internet (40 percent), then television (28 percent), then in magazines (11 percent). Lastly, 53 percent of children noted that the most believable information for school papers or projects can be found on the Internet, followed by books (34 percent), and then people they talk to in person (7 percent). These results are summarized in figure 23.

Overall, children appear to rely fairly heavily on the Internet to access different types of information. It was rated as the most believable source of information for schoolwork, entertainment, and commercial information, as well as second most believable for health information and third most believable for news information.

Some age differences emerged in children's indication of which channels they believe most for specific types of information. For instance, older kids tended to believe entertainment information from the Internet and newspapers more than younger kids did, and entertainment information from books and the radio less than older kids did. Additionally, older children believed health information from the Internet, books, and magazines more than younger children did and health information from the radio less than younger kids did.

With news information, older children believed the Internet, books, and magazines more than younger kids, and in-person and radio sources less than them. For school-related

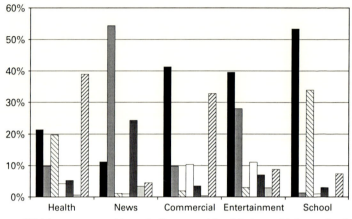

Which source would you believe most for this type of information?

Figure 23

Most credible source by information type

information, older children believed books and magazines more than did younger kids and in-person sources less than them. Finally, the Internet and newspapers are seen as more credible channels for commercial information for older kids than for younger ones, while television was seen as a less credible source of commercial information by older versus younger children.

We next asked children how much people *should* believe the information they find via particular media channels, including newspapers, television, and the Internet. Children indicated significantly different assessments of which medium should be believed, noting that newspapers should be believed the most, followed by television, and finally the Internet, as shown in figure 24. These assessments did not vary with the age of children. Once again, when considered together with the findings above about credible sources by information type, it appears that in some ways children's own use of the Internet may exceed the extent to which they think others should rely on it for credible information.

Methods of Determining Information and Source Credibility

Past research on credibility and on decision making more generally has suggested that there are several ways that people may approach information processing when evaluating information (Metzger 2007; Scott and Bruce 1995). One method is to carefully analyze the information and its features, another is to use a more holistic and intuitive approach based on feelings, and a third method is to draw upon others in one's social circle for advice and guidelines. Indeed, our own research on adults has found evidence for these three strategies in people's credibility

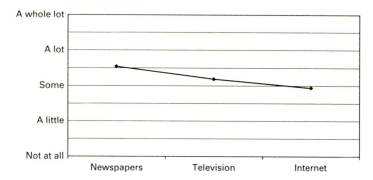

Figure 24
Degree to which others *should* find information credible by medium

determinations, which we call "analytic," "heuristic," and "group-based" (Metzger, Flanagin, and Medders, forthcoming). (See also the next section, "Factors Affecting Children's Credibility Evaluations," for a more detailed description of these methods).

We asked children the extent to which they based their credibility assessments on *heuristic* (e.g., by relying on their gut feelings, making decisions based on feelings, making quick decisions), *analytic* (by carefully considering the information, double checking facts, gathering a lot of information, and considering all views), or *social* (by getting advice from others or asking for others' help) criteria when evaluating whether to believe information online. Kids reported that they used analytic techniques to carefully evaluate the credibility of information online "sometimes" to "often" (3 and 4 on the scale, respectively) whereas they used social and heuristic methods less often overall.

Although this pattern of using analytic methods most often, followed by heuristic and then social methods, was similar across all age groups, the frequency with which kids used each of these strategies increased with age. In other words, there was a general trend in that older children reported applying all three methods of credibility evaluation more often than younger kids (see figure 25).

Interestingly, these results do not comport with research on adults, who indicate that they often use heuristic methods of credibility evaluation. Without further study, however, it is impossible to say whether this difference is due to true differences between kids and adults in their strategies for evaluating

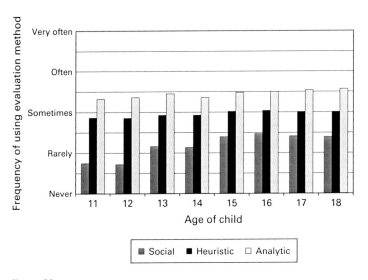

Figure 25
Use of credibility evaluation methods by age

credibility, or to the specific question wording or research method used in this study (i.e., survey methods versus focus groups). Indeed, the question itself may have prompted kids to think about situations in which knowing the credibility of the information they sought was important, rather than considering how they evaluate credibility across the full range of information-seeking situations (e.g., the question asked how often they used analytic, heuristic, or group-based strategies *when deciding what to believe*, rather than simply asking how often each strategy is used while looking at information online).

Moreover, survey responses on items like these are susceptible to social desirability response biases, where study participants want to sound like they are more diligent and informed than they really are. So, while these results are intriguing and even encouraging in that they suggest children are carefully considering the credibility of information they find on the Internet, they should be interpreted with caution until further research can be conducted.

In order to further understand the ways in which children make situation-specific (as opposed to general) judgments about the credibility of information they find online, we asked them how important a number of cues/elements were when they were determining whether to believe information they found online. To do this, each child was asked to imagine a situation in which he or she was seeking information about a certain topic or type of information (i.e., health information, news information, entertainment information, information about something they wanted to buy online, or information for a school paper or project).

Across all five types of information, the most important cues/elements involved the currency of the information, the security of the Web site, information completeness, and the authority of the information source (for example, if the information originated from experts). Next most important were a number of items that dealt with social endorsement and reputation; those cues reported as least important dealt with Web site design and general feelings about the Web site (see table 3 for details).

It is interesting to note that across all of the credibility cues and elements presented, responses ranged from 3 ("somewhat important") to 4 ("important"), and that no cue/element was considered by children on average to be either "very important" or "not at all important." Also, while age differences were not dramatic, there was a slight trend toward many of these cues/ elements to increase in importance for older children. Advances in cognitive development that come with age, accumulated (positive and negative) experiences with online information, or having had information literacy training could explain these results.

Regardless of their explanation, however, the results overall suggest that kids do pay more attention to the "right" cues when determining credibility, at least according to digital literacy advocates and educators who stress the importance of source authority or expertise, security, and information currency and scope as credibility markers (e.g., Alexander and Tate 1999), and that learning to evaluate credibility by examining many facets of Web sites and information provided on them is a process that develops over time.

Table 3

Importance of various credibility cues

Credibility cue	Mean	SD
The information on the Web site is up-to-date	3.85	1.05
The Web site seems safe and secure*	3.80	1.07
The information is very complete	3.71	1.04
Experts believe the information (like your doctor, teacher, etc.)	3.68	1.09
The information is from an expert on the topic	3.66	1.07
The information seems reasonable to you*	3.54	0.99
You ask an expert (like your doctor, teacher, etc.) who you know in person	3.49	1.16
It does not try to convince you to do something or buy something	3.49	1.24
The Web site is easy to use*	3.45	1.16
People you know, such as friends and family, believe the Web site or information source	3.43	1.08
There are high ratings, positive comments, or good reviews	3.41	1.11
Others recommend the Web site or information source	3.39	1.05
You get more than just one person's opinion*	3.38	1.09
You have heard good things about the information source or Web site creator*	3.37	1.11
The information is well written, and you see no typing mistakes	3.30	1.23
The information on the Web site is similar to information on other Web sites	3.28	1.11
There is information about the source's or author's education or training	3.20	1.19
You have heard of the source or information creator before*	3.13	1.14
A lot of other people use the Web site*	3.07	1.17
The information you find is similar to what you already think	3.06	1.10
The Web site address has a certain ending (like ".gov" or ".edu" or ".com")	2.96	1.34

Note: 1 = not at all important, 2 = a little important, 3 = somewhat important, 4 = important, 5 = very important; *this item did not vary by information type

Methods of Determining Information and Source Credibility by Information Type   We also compared the cues and elements that kids use to evaluate the credibility of various types of information individually (i.e., health, news, entertainment, commercial, and school-related information). Regardless of the type of information sought, children reported that the following things were equally important in determining credibility:

Reputation   having heard good things about the information source or Web site creator and having heard of the source or information creator before

Endorsement   getting more than one person's opinion and the fact that a lot of other people use the Web site

Security   the Web site seems safe and secure

Site design   the Web site is easy to use, the Web site looks good, and just liking the Web site

Information plausibility   the information on the site seems reasonable

Differences in the importance of some credibility cues/elements did emerge for different types of information on several items. When it comes to more consequential information, specifically, *health information* and *information for their schoolwork*, kids felt the following cues/elements were particularly important for determining credibility:

Expertise   experts believe the information, the information is from an expert, there is information about the source's education/training (expertise), and they ask an expert whom they know in person

Endorsement   others recommend the Web site

Professionalism   the information is well written and there are not typographical errors

Information currency and comprehensiveness   the information is current and complete

Information consistency   the information is similar to information on other Web sites

Lack of bias   the Web site does not try to sell you something, or the Web site address ends in ".gov" or ".edu"

It is interesting to note that the various expertise items were rated as most important for *health information*, whereas information currency and its completeness, professionalism, consistency with other Web sites, endorsement by others, and lack of commercial motive/bias were rated as most important for *school-related information*. A different picture emerged for *commercial information*, where the most important cues/elements for determining credibility were endorsement (i.e., people you know believe the site or there are high ratings and positive reviews) and similarity of the information to what children already believed.

Nearly all credibility cues and elements were rated as only moderately important for *news information*, except that kids said it was less important for news information to be similar to what they already think or to have high ratings, as compared to most other types of information where they felt these were important factors in credibility assessment. More interestingly, kids felt it was less important that news information be endorsed by others (for example, when people you know believe the Web site or source or when others recommend it), as compared to the other information types. Most surprisingly, children did not rate the importance of news information's completeness and currency as being any higher than for entertainment and e-commerce information, although older kids saw currency of news information as a more important credibility cue than did younger kids.

In almost every case, all credibility cues/elements were rated as least important for deciding whether *entertainment information* was believable. It appears that children recognize that entertainment information is less consequential to evaluate for its credibility compared to the other information types.

In terms of age differences, there was a slight tendency for younger children (11- to 12-year-olds) to feel that most of the cues/elements are less important for evaluating the credibility of each type of information than older children. This may reflect the fact that older kids have had more online experience that would lead them to consider more facets of online information in their credibility assessments than younger kids, or it may reflect simple developmental or experiential differences.

Summary
Overall, kids view information on the Web as relatively credible, and they distinguish across information type and information source when determining the credibility of online information. They show an awareness of the possibility of negative consequences stemming from false information online, and admit that they *should* probably believe some information, such as that on *Wikipedia*, less than they *actually* do. Moreover, children trust information more or less depending on its type,

with some information, such as that used for school projects, seen as more believable than information from strangers they meet online, for example.

Kids also appear to employ different credibility-assessment tools depending on the type of information they are seeking. Our data demonstrate that kids differentiate among news, entertainment, health, commercial, and school-related information when choosing which tools to use. For instance, kids tend to look at fewer credibility cues, and apply these cues with less rigor, when looking for entertainment information than they do when looking for news, health, commercial, or school-related information.

Our study also reveals some slight differences across age groups in the ways kids view the credibility of online information as well as the tools they use to assess information credibility. For instance, older kids are slightly more likely to correctly identify what *Wikipedia* is, and are more likely to use heuristic, social, and analytic methods of assessing credibility simultaneously than are younger kids. Moreover, older kids appear to trust the Internet more as an information source than younger children do, with credibility ratings for entertainment, news, commercial, and health information increasing with age. Older kids (in this case 18-year-olds) also are more likely to report that people should be more concerned about online information credibility than younger kids (12- to14-year-olds).

Interestingly, age differences did not surface in several other areas. For instance, differences do not exist in how much older and younger children believe *Wikipedia* or in how much they thought they should believe it. Children's rating of information credibility across information types also is steady across age groups.

Factors Affecting Children's Credibility Evaluations

In addition to understanding the general parameters of trust among youth, we also analyzed the data to assess what factors play a role in kids' credibility judgments. Surprisingly little research has focused on questions about what types of people are more or less likely to believe information they find online, the extent to which young people exhibit particular Internet usage patterns or online information evaluation practices, or how prior experiences online drive young people's credibility beliefs.

To address these issues, we examined how various factors impact both kids' *credibility concern* as well as their *beliefs about the credibility of Internet information* of various sorts. These factors include:

- demographic or background characteristics
- patterns of Internet usage, access, and experiences
- personality traits
- specific strategies or methods for evaluating credibility

The *demographic or background characteristics* we examined included young people's sex, age, household income, race, and grades in school. Previous research on credibility evaluation has paid scant attention to these sorts of factors, although there is reason to believe that children's information evaluation strategies and opportunities may vary developmentally across age (see Eastin 2008), income (see van Dijk 2006), or by other types of demographic groupings.

We also suspected that certain patterns of *Internet usage, access, and experiences* could impact young people's perceptions of credibility online. In terms of usage, we investigated how much time young people spend with the Internet (per week as well as the number of years they have been using the Internet), their level of skill (technical and search skill, as well as knowledge of the latest online trends and features, as detailed earlier under "Children's Web and Internet Skill Levels" in the section on "Internet Usage among Youth"), and their use of the Internet for specific activities (social networking, contributing information online, visiting virtual worlds, and using the Web for commercial purposes—see the earlier discussion of "Children's Online Activities" in the section on "Internet Usage among Youth").

We also looked at the extent to which children's parents controlled or restricted their access and use of the Internet. To do this, we created a measure that reflected how many restrictions the parent of each child reported imposing in the home, ranging from 0 (parent sets no restrictions on child's use of the Internet) to 4 (parent implements all of the control mechanisms detailed previously under "Parental Control of Children's Access to and Use of the Internet" in the section on "Internet Usage among Youth").

Young people's prior experiences with online information and its evaluation were also included here in order to understand their effect on credibility perceptions. Specifically, we asked children whether they had ever had a bad experience using some information they found online that turned out not to be credible, or whether they had ever heard of this happening to others. We also asked them about whether they have had instruction in evaluating the credibility of Internet information and parents' reports of how often they talk to their children about the trustworthiness of information online (see "Trust and the Negative Consequences of False Information Online" in the previous section, "Perceived Trust and Credibility of Web-based Information").

Several *personality traits* were also explored for their contribution to young people's credibility beliefs and practices, including cognitive dispositions or "thinking styles" that have been shown to influence how people approach information. "Need for cognition," for example, reflects the degree to which people engage in and enjoy thinking deeply about problems or information and, thus, are willing to exert effort to understand and scrutinize information. Another personality trait that we examined, "flexible thinking," measures how willing people are to consider opinions that differ from their own, which we felt might impact the way they process contradictory or contrasting information when judging credibility online. "Faith in intuition" reflects people's tendency to trust based on their first impressions, instincts, and feelings. Survey items were adapted from standard measures of these factors (e.g., Epstein et al. 1996; Kokis et al. 2002) and were pilot tested to ensure that children ages 11 to 18 could comprehend them easily, as noted in the overview section of the earlier chapter, "Research Approach."

A final personality trait examined young people's propensity to trust others, which was measured by questions asking kids how much they feel other people can be trusted, either online or in person (see "Trust and the Negative Consequences of False Information Online" in the previous section, "Perceived Trust and Credibility of Web-based Information").

The last factor we considered was *strategies or methods for evaluating credibility*, or how the *process* of evaluating information online influences the assessments that young information consumers make. We analyzed two sets of strategies for information evaluation. First, we posed the question, "When you decide what information to believe on the Internet, do you . . . [give careful thought to the information, rely on your gut feelings, ask for help from other people, etc.]." Research in cognitive psychology, information processing, and, especially, adolescent decision making (Jacobs and Klaczynski 2005), indicates that adolescents primarily approach information analytically or heuristically when making decisions (Klaczynski 2001). Analytic processing involves effortful and deliberate consideration of information; heuristic decisions are made more quickly, with less cognitive effort and scrutiny. Additional research suggests another decision-making strategy that may be relevant to evaluating credibility online, namely, relying on others to help make decisions (Scott and Bruce 1995). Given the Internet's vast and increasing capacity for social interaction, and the importance of relying on social means of credibility assessment found in previous work on adults (Metzger, Flanagin, and Medders, forthcoming), we included heuristic, analytic, and social or group approaches or strategies of evaluating credibility in our analysis (see also "Methods of Determining Information and Source Credibility," in the previous section on "Perceived Trust and Credibility of Web-based Information").

In addition to these approaches, there exist more specific strategies that young people may use in the context of evaluating the credibility of information online, for example, by focusing more or less on certain credibility cues or elements (see "Methods of Determining Information and Source Credibility"). Further analyses on these various credibility cues showed that they boil down to three strategies: evaluating credibility via social confirmation (e.g., consulting others and looking to see if information is from expert sources), evaluating credibility via information quality (e.g., looking at the currency and completeness of the information), and evaluating credibility via Web site design (e.g., considering the site's appearance and navigability). These constituted the second set of credibility evaluation strategies that we examined to see whether they play a role in young people's credibility perceptions and beliefs.[4]

Concern about Credibility

The degree to which young people are concerned about the credibility of information online is of key interest to this project. Beyond examining levels of young people's concern about whether they can trust the information they find online (which was done in the previous section, "Perceived Trust and Credibility of Web-based Information"), we wanted to understand who is more and less likely to be concerned about credibility, and the degree to which credibility concern is attributable to specific Internet usage and information evaluation patterns.

Toward this end, we looked at how young people's demographic or background characteristics, Internet usage and experience, personality traits, and strategies for evaluating credibility influence their level of concern about credibility online. Analyses showed that the type of strategies that young people use to evaluate credibility affect their concern about credibility. Specifically, those children who are more concerned about the credibility of Internet information tend to use a more analytic than heuristic approach to evaluating information, and rely less on evaluating credibility by means of social confirmation and Web site design. Kids' online experiences and education matter also: having had a bad experience or even hearing about others who have trusted bad information online, having parents talk to them about the trustworthiness of information found online, and having had formal instruction in credibility evaluation all contribute to greater concern about the credibility of information on the Internet.

These results make good sense, since personal or even vicarious negative experiences with online information and formal or informal information literacy training are both likely to sensitize kids to the dangers of using information that is not credible. Also, being more critically minded and thorough in evaluating information may come from a sense of concern about credibility or may contribute to that concern.

The ways in which kids engage with the Internet and participate in content creation also mattered in their concerns about credibility, although less so. More specifically, those who use the Internet to immerse themselves in virtual worlds more often (including playing games such as *World of Warcraft*) and those who contribute information online less, show higher levels of concern about credibility. Also, kids who were more highly skilled and who had been online for a greater number of years were more concerned about credibility. These results indicate that as kids engage more, and more deeply, with various aspects of the Internet, they may develop a healthy sense of skepticism and concern about the believability of information available online. This finding refutes some adults' fears that kids will become more accepting and less critical of Internet information as they deepen their experience and participation in online activities.

Only two traits, flexible thinking style and Internet social trust, emerged as being related to kids' level of credibility concern. As kids are more flexible in considering information that runs counter to their own beliefs and are less trusting of others online, they express greater concern about credibility. Again, this makes sense because attending to contradictory information would naturally raise concern about whose view to trust, as would having little confidence in the trustworthiness of others online.

Interestingly, young people's demographic characteristics did not seem to matter much, with one exception: race made a very small contribution to users' concern about credibility. Kids who reported themselves to be minorities expressed slightly greater concern about credibility than did white children, which may reflect subcultural differences found in many surveys for trust of all sorts among minority populations (Alesina and La Ferrara 2002). It is noteworthy that overall age did not impact concern about credibility (i.e., other factors accounted for differences in credibility concern when considered collectively), despite the fact that older kids have more online experience and more life experience.

Beliefs about the Credibility of Online Information

While young people's concern about the credibility of information online seems to be driven to some extent by analytic processes of evaluating information, this is not the case for their actual trust of online information, both in terms of the amount of information on the Internet they feel is credible and their likelihood of trusting information they personally find online.

Indeed, young people's beliefs about credibility appear to be more a function of heuristic processes, as evidenced by the fact that young people who rated online information as more credible tended to use a more heuristic than analytic approach to evaluating information online. Consistent contributors to young people's actual credibility beliefs were evaluating information based on the Web site's design and using heuristic credibility evaluation strategies, such as relying on gut feelings and making quick credibility judgments. Personality traits related to these heuristic strategies also contributed significantly to beliefs about credibility, whereby youth possessing lower need for cognition and higher faith in intuition thinking styles rated information on the Internet as more credible.

These results are not surprising in light of what is known from past research on adults (see Metzger 2007), which finds a good deal of people's credibility evaluations are based on cursory rather than thorough examination of online information. However, the fact that heuristic processes figure so prominently in how much online information kids find credible and how likely they are to believe the information they find online is a little disconcerting—particularly for digital literacy advocates who stress the need for kids to apply critical thinking skills to

Internet-based information, due to its unique characteristics that make discerning credible from non-credible information particularly complex and difficult (see Metzger, Flanagin, et al. 2003). Another personality trait that influenced young people's views of the credibility of online information was their trusting nature. Questions that tapped into the degree to which kids felt others could be trusted both generally and online were significant and positive predictors of how much of the information online they felt was believable.

Kids' demographic characteristics mattered more for their actual beliefs about the credibility of online information than they did for their concern about credibility. Specifically, young people who were from families of higher income said they believed more information on the Internet, and both younger kids and girls were more likely to believe the information they find online compared to older kids and boys, respectively.[5]

This could be due to differences in girls' and boys' Internet usage or experiences interacting with others online, and to the fact that older children are more likely to have had greater overall exposure to online information generally, and thus perhaps more experiences with bad information, as well as being more likely to have had formal information literacy training than have younger children.

Indeed, the data show that Internet usage and experiences also factor into kids' credibility beliefs. In particular, young people who rated themselves as more technically skilled online felt Internet information was more credible, as did those who use the Internet to visit virtual worlds more often. Past negative experiences with false or non-credible information also mattered in that having such experiences led kids to say that less Internet information is believable and that they were less likely to believe the information they found online, as one would expect.

Beliefs about the Credibility of Other People Online

An area of particular concern among both parents and educators has focused on children's trust of strangers they meet online through online chat groups or forums, social networking sites, virtual communities, and other Internet venues. As discussed earlier (see "Trust and the Negative Consequences of False Information Online" in the previous section), we probed kids about the extent to which they trust people they meet online. Although children did not express a great deal of trust of strangers on the Internet overall, kids' specific uses of the Internet seemed to increase their trust of strangers online, particularly uses such as spending time using the Internet to visit virtual worlds and contributing information (to blogs, personal Web sites, online groups, etc.). While at first glance this may seem troubling, we also found that kids whose parents controlled their Internet access and use to a greater extent showed greater trust of strangers. These findings, coupled with the fact that kids did not express high levels of stranger trust online overall, may indicate that kids who immerse themselves in virtual worlds and contribute online content interact with strangers in spaces that are reasonably safe (or at least parent-approved) and thus they feel they can trust the strangers they meet in those online environments.

It also appears that the strategies kids employ to evaluate credibility contribute to how trusting they are of strangers online. Kids who use more group-based and heuristic credibility evaluation methods, and those who use less analytic methods, trust strangers more. In other words, using more methods to evaluate credibility and being more meticulous in evaluating information leads kids to be more cautious about trusting strangers. Supporting this, the trait of need for cognition also emerged as a factor in online stranger trust, such that kids who were higher in need for cognition expressed that they are less trusting of people they meet online than those with lower need for cognition.

The only other trait that was found to impact stranger trust was kids' general propensity toward trusting others, which is not surprising. Two demographic characteristics, race and age, played a significant but minimal role in the degree to which kids trust people they meet on the Internet. Older kids expressed more trust of strangers, while white kids said they trusted strangers less than did nonwhites.

Of course it is difficult to say exactly why these patterns emerged without knowing more about where kids meet strangers online. Overall, though, our results suggest that there may be relatively little reason for adults to fear kids trusting the strangers they meet online because young people seem to be fairly aware that there are risks to being too trusting of strangers, and because kids can be taught to use more rigorous credibility evaluation strategies that may help increase their acumen for deciding not only what, but whom, to trust online.

Beliefs about the Credibility of Blogs and *Wikipedia* as Information Sources

As relatively new sources of online information that rely heavily on the contributed knowledge of largely unknown others, we were interested in understanding what drives young people's credibility perceptions of blogs and *Wikipedia*.

Our analyses showed that, despite kids feeling that the information in blogs is not as credible as news and political information in newspapers and television news overall, kids who used fewer analytic and more group credibility evaluation strategies (and who relied on Web site design more in their evaluation practices), who felt more strongly that people could be trusted online generally, whose parents talk with them more often about the credibility of online information, and who were from households with lower incomes were more likely to find blogs credible than kids who did not possess these characteristics.

Among kids who have some familiarity with *Wikipedia*, those who express greater trust of others online, employ more heuristic credibility evaluation strategies and rely more on site design to discern credibility, and who use the Internet more frequently for visiting virtual worlds, believe *Wikipedia* to be more credible than kids who do not. Grades were the sole demographic factor at play here, where kids with higher grades rated *Wikipedia* as more believable than did kids earning lower grades in school. Finally, kids who have had a bad experience themselves (or who personally know others who have had bad experiences) by trusting information that later turned out to be false find *Wikipedia* to be less believable compared to young people who have not had these negative online experiences.

While these results show that kids' perceptions of the credibility of blogs and *Wikipedia* are each affected by some unique factors, both show that heuristic processes or strategies of determining credibility, as well as individuals' propensity to trust others online, are prominent drivers of children's credibility judgments of these two relatively new Web-based information resources.

Factors Contributing to Children's Credibility Evaluation Methods

Because the methods by which kids evaluated credibility online emerged as a significant factor in young people's credibility attitudes and judgments across several of the aforementioned analyses, we wanted to understand better what leads to particular styles or methods of credibility assessment. To do this, we explored what predicted whether children used a more analytic, heuristic, or group-based (i.e., social) strategy when evaluating online information credibility.

As discussed in the introduction to this section, using an analytic strategy to evaluate credibility is somewhat effortful, since it involves careful inspection of the information and its author or source (e.g., gathering a lot of information and double checking it). Young people who tend to use this strategy exhibit a number of related behaviors and traits; for example, they enjoy thinking hard about things, are more flexible thinkers, get higher grades in school, and have more technical Internet skill. They are also older and look more to the quality of the information and less to site design when evaluating the credibility of information online. In other words, a pattern of analytic evaluation appears to be related to individuals' intellectual prowess and experience, which comes as no surprise. Other factors that predict whether someone uses analytic evaluation strategies include skepticism about whether others online can be trusted, using the Internet for commercial purposes (which may breed skepticism), and being from lower, rather than higher, income households.

Heuristic credibility evaluation strategies tend to be based on emotion and are made quickly, without much consideration of evidence, sources, or information. Major predictors of this style of evaluation tended to fall into the category of personality traits. For example, the trait of trusting one's instincts or hunches and going by one's gut feelings to evaluate information were the most significant factors predicting heuristic evaluation strategies. Related to this, using Web site design cues to guide credibility decisions, and *not* relying on information provided by experts were also important. Another trait, need for cognition, negatively predicted using a heuristic strategy, such that those who avoid thinking hard about problems and do not enjoy activities that are cognitively demanding used this strategy more. Finally, individuals' trust of others online increased their tendency to evaluate credibility heuristically.

The last strategy we examined was using group-based methods of evaluating credibility (i.e., seeking the advice of others to help discern whether some information or person online is credible). Here, various patterns of Internet usage and experience were the best predictors. Contributing information online, engaging in social exchange, and using the Internet for "virtual" purposes all resulted in relying to a greater extent on others to help discern credible information online. This makes sense, of course, as kids who participate in these sorts of online activities would be likely to have a larger and more (inter)active social circle that they can turn to for advice and guidance while online.

Group-based credibility assessments were also predicted by greater technical skill and trust of others online, as well as by more life experience (age) and years of Internet experience. Moreover, kids who indicated they used group-based strategies were more likely to look at Web site design and to rely on social confirmation, and they were less likely to consider the quality of the information in determining credibility. Children using social evaluation strategies were also less disposed to relying on gut feelings for trust in their general lives.

Summary

As we suspected, data show that young people's uses of the Internet, demographic and personality traits, and specific strategies for evaluating credibility all played a role in the judgments that children make about whether to trust information and people they encounter online.

Among the more consistent predictors of both credibility concern and beliefs were the personality traits of need for cognition and kids' propensity to trust others, either offline or online. Online experiences also mattered, as credibility beliefs seem to be shaped by past negative experiences with finding and using bad information. The old adage of "once bitten, twice shy" definitely seemed to operate in this arena.

Usage was important too, and in particular using the Internet to immerse oneself in virtual environments (i.e., playing online games, participating in online communities) was associated not only with greater concern about credibility, but also with stronger beliefs that online information is credible. This suggests that, contrary to some views, online gamers and participants in virtual worlds have an optimistic yet appropriately skeptical sense of the credibility of information and others online.

The most consistent influence on young people's credibility perceptions was the strategies kids used to evaluate information (which were themselves influenced by a number of the other credibility predictors, such as traits like need for cognition). Whether kids used analytic, heuristic or, in a few cases, group-based credibility assessments impacted their credibility concern, as well as their beliefs about credibility. Interestingly, although concern about credibility appears to be driven by analytic credibility evaluation processes, actual beliefs about the credibility of information found is dictated by more heuristic processes. This suggests that while most kids take the idea that they should be concerned about credibility seriously (by invoking a systematic and analytical approach), those who feel the Web is more credible arrive at that judgment by taking a more lax approach in actually evaluating the information they find online.

Finally, it is interesting that demographic characteristics did not emerge as particularly important or consistent predictors. This suggests that, at least in the area of credibility perceptions, there is little evidence of a digital divide. Similarly, Internet skill and frequency/duration of Internet usage were minimal factors in kids' credibility beliefs, as was instruction in information literacy practices.

Child/Parent Dyads and Credibility Assessments

Past research has demonstrated that an individual tends to feel that he or she is less susceptible to negative influence than are other people. This phenomenon has its roots in a cognitive process known as the "optimistic bias" (Weinstein 1980), which is the tendency to see oneself as less likely than others to experience negative life events (or, conversely, more likely to experience positive life events).

Research on the notion of optimistic bias has thoroughly examined its impact on the beliefs and behaviors of individuals in both health (Clarke et al. 2000; Weinstein 1982) and non-health (Weinstein 1980) contexts, and has demonstrated the stability of this phenomenon across a wide range of demographic variables, including age, sex, and education (Weinstein 1987). However, little research has focused on the occurrence of the optimistic bias in a digital media environment (for an exception, see Campbell et al. 2007) and research to date has not examined child-parent dyads with regard to this phenomenon.

To fill this void, we were interested in seeing if the same psychological processes underpinning the optimistic bias phenomenon operate in the context of judging the credibility of information online. We were particularly interested in children's perceptions of their own ability to evaluate the credibility of information online, and to differentiate between good and bad information compared to (a) their parents and to (b) a typical Internet user. For children, survey questions designed to assess this were as follows:

• Who is more likely to believe false information on the Internet, you or your parent?

• Who is more likely to question the information they find on the Internet, you or your parent?

• Who is better at figuring out which information is good or bad on the Internet, you or your parent?

• Who is more likely to believe false information on the Internet, you or a typical Internet user?

• Who is more likely to question the information they find on the Internet, you or a typical Internet user?

• Who is better at figuring out which information is good or bad on the Internet, you or a typical Internet user?

Moreover, we took advantage of our study design to query parents and their children to look at whether parents also exhibited an optimistic bias. To do so we posed a parallel set of six questions that compared parents' perceptions of their own ability to discern credible information online to that of (a) their child, and (b) a typical Internet user.

Our data on the optimistic bias are unique in that they are the first to examine this phenomenon in the context of credibility judgments in digital media environments, as well as to provide a detailed look at the phenomenon in the context of parents and children, and not just between individuals and "typical others."

Children's and Parents' Comparisons to Each Other
Figure 26 shows that across all ages children consistently feel that their parents are better than they are at figuring out which

information is good or bad online. Similarly, parents seem to agree, as they also consistently reported that they were better than their children at discerning good from bad information on the Internet. So, while the parents' overall responses display evidence of an optimistic bias, the children's responses do not.

There is an interesting trend with age, however. As children get older, the gap between their perceptions of their own and their parent's ability to differentiate between good and bad information online narrows. So, while younger children feel their parents are significantly better at discerning credible information, by about age 16 they feel they are about equally good as their parents at doing so. Interestingly, while parents do

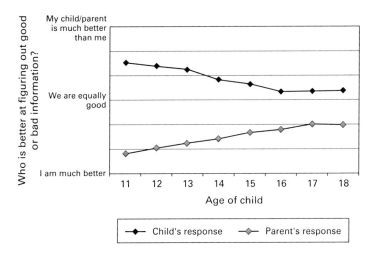

Figure 26
Child/parent differences in discerning good versus bad information online

perceive their children's ability to tell good from bad information online as improving with age, they never report feeling
their children are their equal in this regard.

When it comes to who is more likely to believe false information online, we see a similar pattern of responses, such that
parents always feel their children are more likely to believe
false information than they are. Younger children feel that they
are slightly more likely than their parents to believe false information, but again by age 16 kids feel they are about as likely
as their parents to be tricked by false information on the Internet (see figure 27). Notably, at age 16 children actually cross
over the midpoint of the scale, indicating that on average
they feel they are less likely than their parents to believe false

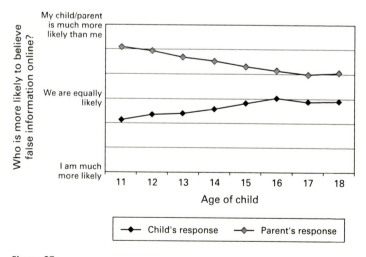

Figure 27
Child/parent differences in discerning false information online

information online. On average, 17- and 18-year-olds also approach this threshold quite closely.

When asked about who is more likely to question information they find online, a slightly different pattern emerges. Figure 28 shows that while parents still consistently believe they are more likely than their children to question information found online—and children seem to agree (although not as strongly)—the lines do not change as dramatically across the age groups as they did for the other comparison questions.

This indicates that both parents' and kids' perceptions of how likely they are to question information online stay fairly consistent across the age groups, with only slight upward and downward trends for the parents and their children, respectively.

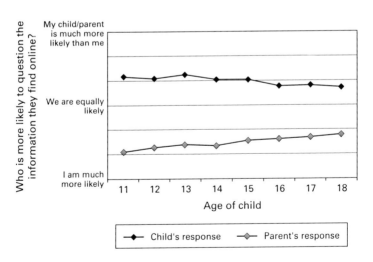

Figure 28
Child/parent differences in the likelihood of questioning information online

This could suggest a generational difference in the level of skepticism by which parents (so-called digital immigrants) and children (so-called digital natives) approach online information, or at least a difference in the amount of effort that each group is willing to expend in thinking about the credibility of the information they find. Effort, of course, may be tied to the type of information sought online, with the idea that people will expend more energy questioning information that is more personally and objectively consequential (e.g., health, financial information). Moreover, it is likely that adults spend more time overall while searching online for these types of information than do children.

Children's and Parents' Comparisons to Typical Internet Users

In comparison to a typical Internet user, even the youngest children saw themselves as slightly better on average in their ability to figure out which information is good and bad online, thus showing an optimistic bias. Older children viewed themselves as even more capable than a typical Internet user in this regard (see figure 29).

Parents of children across all age groups also felt they were better than the typical Internet user in discerning credible information online. That said, by age 17 kids' and their parents' ratings of their own ability are very similar to each other, and they both see themselves as more skilled than typical Internet users.

Parents and children across all ages felt that they were less likely than a typical Internet user to believe false information online, again showing clear evidence of an optimistic bias for both groups. In addition, parents always view themselves as

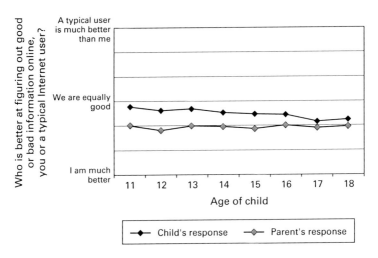

Figure 29

Child/parent differences in discerning good versus bad information online, compared to a typical Internet user

more capable in this regard than their children. Interestingly,
older kids (ages 15 and up) and their parents were both slightly
more likely than younger kids and their parents to report a
favorable comparison to typical Internet users when it came to
believing false information online, as can be seen in figure 30.

An optimistic bias was also seen for both groups, although
more strongly for parents, when they were asked whether they
or a typical Internet user were more likely to question information found on the Internet. As seen in figure 31, parents consistently rated themselves as more likely to question information
than a typical Internet user, as do their children, and there is an
increasing tendency to do so as kids get older. Again, parents
always view themselves as more capable than their children.

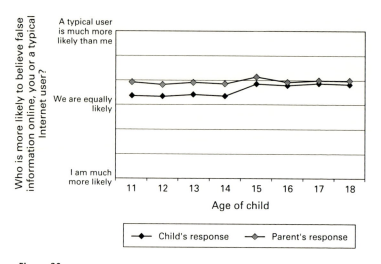

Figure 30
Child/parent differences in the likelihood of believing false information online, compared to a typical Internet user

Summary

Overall, compelling evidence exists for an optimistic bias in individuals' perceptions of their ability to evaluate the credibility of information online, especially on the part of parents. Clear evidence for an optimistic bias exists across several measures, for parents compared to their children and for both parents and children compared to a typical Internet user.

Although compared to their children parents might be realistic in their estimations, it is unlikely that, as a group, the parents and kids surveyed are uniformly more accurate in their judgments of credible information online than are typical Internet users. Our data are consistent with some of the preliminary trends that have emerged from research into the optimistic bias

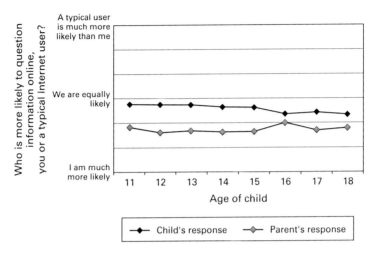

Figure 31
Child/parent differences in the likelihood of questioning information online, compared to a typical Internet user

in a digital media context, which has shown that people believe themselves to be less likely to have their credit card or identity stolen online, and more likely to avoid being misled by information online, as compared to their peers (Campbell et al. 2007).

That said, it is clear that more research into this phenomenon needs to be conducted, in order to solidify the trends that have emerged in this study and to better understand how the optimistic bias operates in an environment in which much of our information about others is produced and consumed electronically. However, our results serve as an important and comprehensive first step in this endeavor, offering a valuable look at the nuances of the optimistic bias both within and across generations.

Web Site Exposure and Evaluation

In order to simulate children's Web-browsing experiences, and to evaluate their reactions to specific Web content, we included a quasi-experimental component in our survey. We presented each child with two stimuli, one at a time: first, they viewed a screenshot of a Web page, which was presented as a "picture of a Web page from the Internet." Depending on the experimental condition to which they were randomly assigned, the child saw either an image from an online encyclopedia or from Amazon .com, followed by questions about the site they had seen (as described in greater detail later). Second, children were presented with another Web page screenshot, this time of a "hoax" site that currently exists on the Web. Children were again asked some questions about the site they viewed. To avoid any influence on our respondents from other questions on the survey dealing with credibility that might cue them to this concern, this portion of the survey preceded all other sections.

Online Encyclopedia Exposure and Reactions

A critical feature of the contemporary Internet environment is the ability of users to be both information consumers and information providers. Indeed, the Internet's very design facilitates wide-scale collaboration among individual users (Flanagin, Flanagin, and Flanagin 2010), which can take a number of forms, ranging from the provision of valuable consumer information to the organization of political protests. Such collaborative efforts often rely on individuals pooling their efforts to create collectively held resources that none could produce without the aid of others (see, for example, Benkler 2006 and Jenkins 2006).

One venue in which collectively produced information has burgeoned is in online encyclopedias such as *Wikipedia*, where anyone can anonymously contribute encyclopedia entries or edit those provided by others. As mentioned in an earlier section ("Perceived Trust and Credibility of Web-based Information"), in its short history *Wikipedia* has risen to among the top 10 most popular Web sites in the United States today, with more than 3 million user-generated entries (Alexa 2009; Quantcast 2009).

Given that all of the content in *Wikipedia* is provided by anonymous individuals, there has been a great deal of controversy and concern regarding the credibility of this information, particularly as compared to more established encyclopedias such as *Encyclopaedia Britannica*, which also has an online presence. In addition, exclusively online encyclopedias like *Citizendium* have also emerged, relying on user-contributed content that is not anonymous and that is provided or vetted by experts prior to its acceptance.

In spite of concerns that user-generated content may be less credible than its expert-produced counterparts, studies suggest that the differences may not be particularly great. For example, research has shown that user-created entries in *Wikipedia* have been judged to be nearly as accurate as well-regarded print encyclopedias like *Encyclopaedia Britannica* (Giles 2005), and entries from *Wikipedia* have been evaluated as relatively credible, and even more so by area experts than by non-experts (Chesney 2006).

In light of the relatively heavy use of and reliance on *Wikipedia* by children (e.g., 84 percent have used it to look up information, as detailed in the previous section), and the fact that no research to date has systematically considered children's use of online encyclopedias, we endeavored to assess children's perceptions of the credibility of information in online encyclopedias. To do so, we performed several different quasi-experiments.

In the first, we had a subset of children who took the survey view a screenshot of an encyclopedia entry, which was presented as coming from one of three different online encyclopedias. In reality, encyclopedia entries were actually identical in all cases within each experimental condition, and were derived from information gleaned from all three of the sources.

The notable difference among the encyclopedias was the purported source of the information, which was reflected in the description of the encyclopedia that children were given: children were instructed that they would see a picture of a Web page from (a) "the online encyclopedia *Wikipedia*, where anyone can add or change information at any time without giving their real names," or from (b) "the online encyclopedia *Citizendium*, where anyone can contribute entries, as long as they are identified by their real names. All contributions, however, are reviewed by experts before being accepted," or from (c) the online version of "*Encyclopaedia Britannica*, whose entries have been contributed by respected experts worldwide since 1768." To ensure that children in this study understood these differences, they were asked to later identify which method of selection for entries was used by the encyclopedia they viewed. Those children who did not correctly identify the method of selecting entries for the encyclopedia (41 percent of respondents) were excluded from all further analyses.

To assess differences across encyclopedia entry topic, three different types of entries were provided and were assigned randomly (an entry on an entertainment topic, an entry on a news topic, and an entry on a health topic). Tests showed no differences across encyclopedia entry topic, so these entries were collapsed for subsequent analyses.

Encyclopedia entries were also constructed to be either one-sided in their presentation of information or balanced in their presentation. Pretests with a different sample of young adults confirmed that stories were perceived as appropriately one-sided or balanced. Data from the children in the present study confirmed this as well. Figure 32 shows an example encyclopedia entry, presented as a balanced entry on the topic of autism originating from *Encyclopaedia Britannica*.

Results from this quasi-experiment yielded two interesting discoveries. First, children found the balanced encyclopedia stories to be significantly more believable than the one-sided version of the same stories, as illustrated in figure 33. Second, children found the entries that they believed had originated from *Encyclopaedia Britannica* to be significantly more believable than those they believed originated either from *Wikipedia* or *Citizendium*. However, children did not distinguish between encyclopedia entries they believed originated from *Wikipedia* or *Citizendium*, in terms of how believable they thought they were. Figure 34 shows this relationship. The relative balance of stories

Autism, causes

Main

Autism is a brain development disorder characterized by impaired social interaction and communication, and by restricted and repetitive behavior. Symptoms typically begin before a child is three years old.

It has long been presumed that autism has a strong genetic basis, although the genetics of autism are complex and many autistic individuals have family members who are unaffected by the disorder. There is increasing suspicion that autism is instead a complex disorder with several causes that may co-occur. One proposed cause is thiomersal, which is a mercury-based compound used as a preservative in childhood vaccines since the 1930s.

Some feel mercury may contribute to the development of autism and other brain development disorders in children, and in July 1999, following a review of mercury-containing food and drugs, the Centers for Disease Control (CDC) and the American Academy of Pediatrics asked vaccine makers to remove thiomersal from vaccines as quickly as possible. It was rapidly phased out of most U.S. and European vaccines.

However, the removal of thiomersal coincided with statements from scientific bodies indicating that it was harmless, sparking confusion and controversy. Thousands of lawsuits were filed in the U.S. to seek damages from alleged toxicity from vaccines, including those with thiomersal. Parents were especially concerned about the disorder's rapid growth since the late 1980s, and a study at the University of Texas demonstrated a link between mercury exposure and autism, finding a relationship between industrial release of mercury and community autism rates. Others note the similarity between the symptoms of mercury poisoning and those of autism.

Scientific and medical bodies, including the Institute of Medicine, the World Health Organization, the Food and Drug Administration, and the CDC, however, reject any role for thiomersal in autism. Multiple lines of scientific evidence have been cited to support this conclusion. Most conclusively, eight major studies as of 2008 demonstrated that autism rates failed to decline despite removal of thiomersal from vaccines, arguing strongly against a causative role. Research on autism and mercury is still on-going.

Figure 32

Example encyclopedia entry

and their source did not demonstrate any statistically significant interaction effects.

These findings indicate that children readily attended both to the information content in the entries and to the source of the information when asked to evaluate its credibility. Content was important inasmuch as children rated the balanced stories as more credible than one-sided presentations, consistent with past studies that demonstrate higher perceptions of credibility for more balanced presentations, at least within some contexts

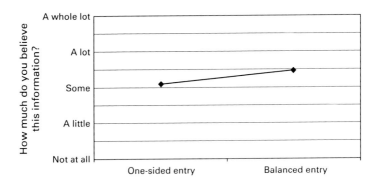

Figure 33
Credibility of one-sided versus balanced encyclopedia entries

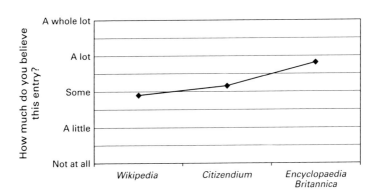

Figure 34
Credibility of encyclopedia entry by perceived source

(Kamins et al. 1989). Information source was also important, confirming past research that has demonstrated differences in credibility based on the perceived information source (Flanagin and Metzger 2007). However, it is unclear whether any preconceived notions they had about the three sources in this study attenuated or exacerbated the effect of the source above and beyond that which our descriptions were designed to provoke.

We next endeavored to assess whether encyclopedia entries that *actually* originated from these various online sources (as opposed to those we created) were viewed differently among children with regard to their credibility. We also evaluated whether it made a difference from which among the three online encyclopedias children believed the entry to have originated.

To do this we selected actual entries on two different topics (global warming and homeopathy) from each of the three online encyclopedia Web sites and edited them very slightly to be of roughly the same length (content was not changed). Tests once again showed no differences in believability across encyclopedia entry topic, so data from these entries were collapsed for subsequent analyses.

We showed children a screenshot of one encyclopedia entry, presented as if it originated from one of the three encyclopedias. However, the encyclopedia entry may have actually originated from any of the three encyclopedias. In this manner, we created 18 different page images, representing each possible combination of encyclopedia entry *topic* (global warming or homeopathy), original encyclopedia entry *source* (*Wikipedia*, *Citizendium*, or *Encyclopaedia Britannica*), and the *placement* of the encyclopedia entry (*Wikipedia*, *Citizendium*, or *Encyclopaedia*

Britannica). Figure 35 shows an example of an entry from *Wikipedia* on the topic of global warming, although the text for this entry actually came from *Citizendium*. Again, only those children who correctly identified the encyclopedia's actual method of selecting entries for the encyclopedia were included in subsequent analyses.

Results showed that, by itself, where the entry actually originated (i.e., the original and actual source of the entry) was irrelevant to how believable the entry was found to be by children. Thus, the *source* of the encyclopedia entry was not important

Figure 35
Example encyclopedia entry

with regard to its perceived believability. The *placement* of the entry, however, was critical in children's credibility evaluations: encyclopedia entries were assessed as significantly *less* believable when placed on *Wikipedia*'s site than when they were placed on either *Citizendium*'s or *Encyclopaedia Britannica*'s sites (and children did not report statistically significant differences between these two sites), as illustrated in figure 36.

Moreover, the entry placement also interacted in meaningful ways with the entry source, such that, for example, entries actually originating from *Wikipedia* were perceived as significantly *more* believable when they appeared on *Citizendium*'s Web page than if they appeared on *Wikipedia*'s page, and even *more* believable if they appeared to have originated from *Encyclopaedia Britannica*. Put another way, the encyclopedia entries from *Wikipedia* were seen as significantly more believable than those

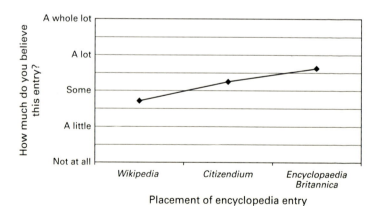

Figure 36
Credibility of encyclopedia entry by placement of entry

from both *Citizendium* and *Encyclopaedia Britannica*, but only when children thought they were *actually* from *Citizendium* or *Encyclopaedia Britannica*. Figure 37 illustrates these results.

Similar to the previous encyclopedia quasi-experiment, children in this case show strong evidence of attending carefully to the entry (in this case, its placement and apparent source). Interestingly, the fact that *Wikipedia* content was deemed more credible if children thought it originated from *Citizendium*, and most credible under the banner of *Encyclopaedia Britannica*, could be taken as signaling the high quality of *Wikipedia* information, despite popular cries that it cannot be of high credibility since it is provided by anyone who cares to contribute it.

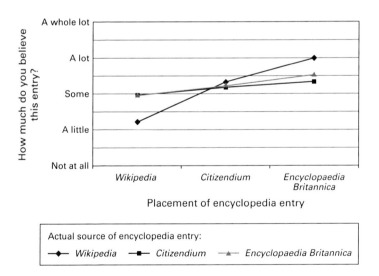

Figure 37
Credibility of encyclopedia entry by actual encyclopedia source

Commercial Web Site Exposure and Reactions

People are increasingly relying on the Internet for commercial information and e-commerce transactions that range from small personal items to home purchases. Although due to financial constraints and other factors children constitute only a small proportion of consumers online, the knowledge and habits learned as children are likely to influence their use of the Internet in this capacity well into adulthood. It is therefore important to understand their perceptions of the credibility of commercial Web site information and the factors they find important in their evaluative processes.

Overall, children's consumption of commercial information online is low, and they do not find it to be very credible as compared to other types of information. Yet, kids believe that commercial information is better retrieved online than via other sources. Moreover, both their use of commercial information and their faith in its credibility increase with age. Finally, they tend to rely on product endorsements by other Internet users when making purchasing decisions, suggesting the need to assess their facility in evaluating such cues in relatively naturalistic commercial environments (see the previous sections of this report for more complete discussion of these findings).

To gauge the degree to which children make credibility assessments of commercial information online, and the factors that influence their evaluations, we presented a subset of the children surveyed with one screenshot from a set of product pages from Amazon.com, which were slightly modified in order to highlight particular features available on these pages. Three different products were shown (a digital camera, an electric toothbrush, and rolling luggage), in order to determine if the type of product in question influenced young people's assessments of information credibility and product quality.

Given recent attention in studies of Web-based information credibility to the influence of others' opinions in credibility assessments (Flanagin and Metzger 2008; Metzger, Flanagin, and Medders, forthcoming), we focused on the prevalence and nature of user-generated feedback in forming children's assessments of commercial products. Specifically, we varied (a) the *number* of ratings and (b) the *average* rating provided about products by other users, by altering this information on the Web page screenshots used in the study. The pages thus showed the number of user ratings as 4, 16, 102, or 1,002 and average "star" ratings (on a 1–5 scale, where 5 is the best rating) of 1.6, 2.23, 3.0, 3.68, 4.4, 4.84, or 5.0. In this manner, we created 84 different page images, representing each possible combination of number of ratings, average ratings, and product. Factors other than these were held constant across all pages. Because we found that children's interest in each product varied, we statistically controlled for interest in the product in all analyses. An example page viewed by children in the study is shown in figure 38, demonstrating a digital camera presented as receiving an average user rating of 3.68, across 102 individual user ratings.

Overall, children found user-provided commercial information to be credible and important, demonstrated by the facts that (a) their assessments of product quality and (b) their likelihood of buying the product in question depended on the product ratings they viewed. There were, however, distinct effects for the number and nature of product ratings: specifically, there

Figure 38
Example product Web page

were strong differences in children's assessment of product quality and likelihood of purchasing the product based on the *average* ratings of the product, but only extremely minor differences based on the *number* of ratings the product received. Thus, although the average ratings positively influenced children's product quality assessments and their likelihood of purchasing the product, the number of ratings for the most part did not. This difference is illustrated in figure 39, which shows the average perceived product quality by the number of ratings, and figure 40, which demonstrates the average perceived product quality by the average product rating. Age was not a major factor in these findings, although there was some evidence that older children were more influenced by the *combination* of average ratings and the number of ratings together.

Although ratings are clearly credibility cues affecting children's product quality assessments and intent to purchase, other factors are reported to be even more important as children look for things to buy online. As illustrated in figure 41, ratings are seen by children as less important than a product's price and product details, but more important than who makes or sells the product.

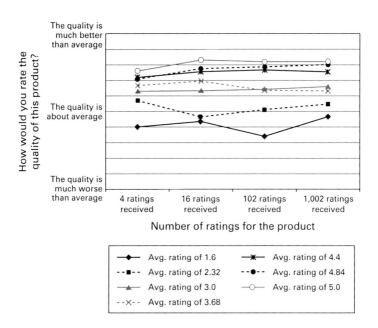

Figure 39
Product quality by number of ratings, across average product ratings

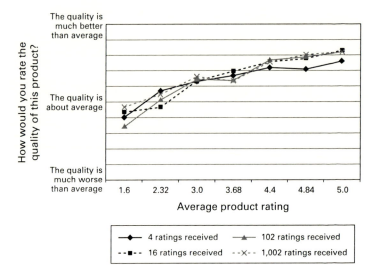

Figure 40
Product quality by average product ratings, across number of ratings

As with the encyclopedia quasi-experiments, children dem-onstrated an ability to attend to specific aspects of the informa-tion available to them that then played a major role in determining credibility and had an influence on their attitudes. In this case, the average rating for a product seemed to override concern for the number of people who rated it (though there was limited evidence that older children were slightly less prone to this), a potentially detrimental oversight given the question-able accuracy of such ratings under circumstances when a small number of people provide feedback. For example, one disgrun-tled consumer has a very large impact on the overall rating of a product when there are only 4 ratings and a very small impact

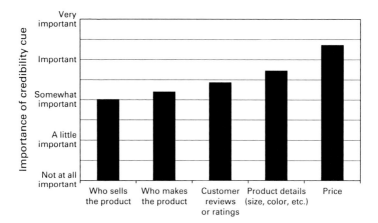

Figure 41
Importance of credibility cues for purchasing decisions

when there are over 1,000, so decoupling the number of ratings from the average rating, as the children in our survey did, reflects a critical deficiency in young people's ability to correctly interpret the meaning of online ratings.

Hoax Web Site Exposure and Reactions

Research (see Krane 2006; Leu et al. 2008) has shown that even among seventh-graders (typically 12 years of age or so) identified as their schools' most proficient online readers, many fail to distinguish fake from legitimate information online. In a study of 25 such students, for example, all believed the information on a hoax Web site advocating the protection of the Pacific Northwest tree octopus, with 96 percent of the students rating the site as "very credible" and recommending the site to others. Moreover, even after learning that the site was a fake, these children had difficulty indicating why this was the case, in spite of clear cues present on the site.

To further assess the degree to which children believe fictitious information online, we presented children with one of two "hoax" Web sites[6] currently present on the Internet. The first site detailed "the first male pregnancy," and included information about the pregnant individual and testimonials and links to media coverage of the pregnancy (see figure 42). The

Figure 42
Male pregnancy hoax Web site

Figure 43
Rennets hoax Web site

second hoax site was devoted to a cause to "save the rennets,"
described as "small hamster like rodents" who are used in the
production of cheese (see figure 43). Tests showed that these
two hoax sites did not differ in terms of their believability,
although there were some other minor differences between
them (as detailed later.)

Six percent of children reported having seen the male pregnancy site before the study and 5 percent reported they had
seen the rennet site previously. After removing these children
from subsequent analyses, an additional 2 percent for the male
pregnancy site and 7 percent for the rennet site indicated, when
asked, that they had taken time out from responding to the
survey to search the Web for information on the sites. Although
these children's responses were also removed from further analyses, this is an intrinsically interesting result in itself, since it
demonstrates children's active use of the Internet to verify
information whose credibility may on its surface be suspect.
Unclear, however, is whether or not this bit of additional information seeking produced lower or higher levels of trust in the
Web site. Neither age nor sex of the child was indicative of the
likelihood to seek out additional information on these hoax
sites.

Overall, and somewhat unlike past studies, children were relatively *unlikely* to believe the information on these hoax sites.
For the male pregnancy site, the average believability score was
1.97 (on a 5-point scale), corresponding to an indication that
children believed the information on the site "a little bit" (see
figure 44). Moreover, 48 percent found the information to be
"not at all believable" and an additional 41 percent reported
that they believed the information either "a little bit" or "some."
Eleven percent of children, however, reported believing the
information either "a lot" or "a whole lot." When asked about
particular features of this hoax site that might inform their
assessments, children on average tended to mildly disagree with
statements that the information on the site was reasonable,

authoritative, well written, and similar to their own beliefs. Also, they tended on average to disagree that the site looked good and that there was evidence on the site supporting the claims made.

Results were similar for the rennet site, where the average believability score was 1.93, indicating that children believed the information on the site "a little bit." Forty-nine percent found the information to be "not at all believable" and an additional 39 percent reported that they believed the information "a little bit" or "some." Eleven percent of children did report believing the information either "a lot" or "a whole lot," however. By and large, children on average tended to mildly disagree with statements that the information on the site was reasonable, authoritative, and based on evidence, even though they tended to find the information to be somewhat well written and

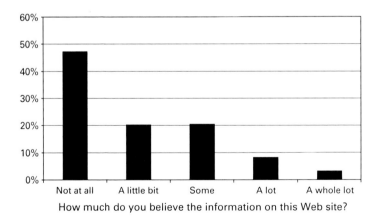

How much do you believe the information on this Web site?

Figure 44
Credibility of male pregnancy hoax Web site

the site to look good. On average, they found that the information on the site was not similar to their own beliefs. Figure 45
shows the extent to which children found the site to be
believable.

Finally, we were interested to know whether there were certain background and demographic characteristics, Internet
usage and experience patterns, personality traits, or methods of
evaluating credibility that were related to believing the hoax
sites.

We found no differences in income, race, or school grades
among children on how believable they found the information
on either of the hoax Web sites to be. However, females and
younger children were more likely to believe the hoax sites.
More specifically, 11-year-olds were significantly more likely to

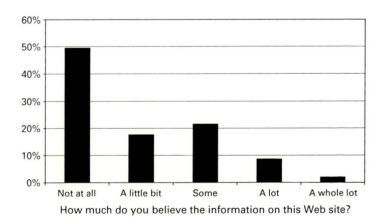

How much do you believe the information on this Web site?

Figure 45
Credibility of rennets hoax Web site

believe information on the rennet hoax site than were 15- to 18-year-olds.

Credibility-evaluation strategies were also related, such that those who used more heuristic methods of credibility evaluation (i.e., rely on site design) found the hoax cites to be more credible than those who used analytic methods of evaluating online information.

Two surprising results were: (a) greater self-reported skill in using the Internet and (b) having had information literacy training by a teacher or parent resulted in higher believability of the hoax sites. In some ways, this is consistent with the work of Leu et al. (2008) who found that even high-performing online readers who had formal information literacy training had trouble discerning whether a hoax site was credible. This suggests that a new strategy for digital literacy training may be necessary.

Results from this quasi-experiment are in many ways quite heartening, although there remain indications for some concern. A majority of children displayed an appropriate level of skepticism when presented with either of the hoax sites, a trend that contradicts prior research on this type of site. Additionally, they seemed able to identify important credibility cues on these Web sites, such as whether information was reasonable, authoritative, and well-evidenced. Perhaps most important, and contrary to past experiments looking at information verification behaviors (Flanagin and Metzger 2007), a number of respondents even reported searching the Internet for more information on these sites after initial exposure to the site. This demonstrates a familiarity and comfort with information-search strategies and credibility-assessment behaviors that is hoped for from digital natives. Nonetheless, approximately 10 percent of children still believed these hoax sites either "a lot" or "a whole lot," indicating some lingering and important concerns, especially for younger Internet users.

Summary

This series of quasi-experiments was designed to place children to some degree in the kind of information environment they might reasonably be expected to occupy during their time on the Internet. Children were tested on their ability to detect both good and bad information, and for the most part seemed able to do so, across both informational and commercial contexts. Their tendency to overlook some of the nuances of information presented on the Amazon.com pages might be explained as the byproduct of a general unfamiliarity with that context, as children do not typically have the resources necessary to engage in a large number or variety of e-commerce transactions. And, they seemed able (at least implicitly) to pick up on the nuances of balanced versus one-sided information in the encyclopedia entries, as well as important information-sourcing cues. Finally, children were (mostly) successful in seeing through the hoax sites they encountered. Although such sites make up a minority of the Internet, the skills children reported drawing upon to handle them are universally applicable.

Conclusions and Implications

This report describes in detail how the activities that young people between the ages of 11 and 18 engage in online, as well as a number of their traits and attitudes, affect their assessments of the credibility of information, and how they go about forming those assessments. Results described herein are generalizable to households in the United States with Internet access.

Summary

The research outlined in this study describes youth who have been using the Internet for much of their lives and who use it for a wide variety of purposes. In many ways, the results are encouraging. For example, the young people in our survey demonstrated an understanding of the potential negative consequences of believing false information online, a tendency to question information that comes from deceptive sources like hoax Web sites, the ability to differentiate between one-sided and two-sided information presentations, general feelings of distrust toward strangers on the Internet, and the inclination to

put more effort into assessing the credibility of highly consequential information (e.g., health information) than less
consequential information (e.g., entertainment information).
Moreover, parents, teachers, and others provide children with
some level of guidance and training with regard to issues of
credibility. Consequently, worries that all adolescents are help-
less and at the mercy of unscrupulous others on the Internet
appear to be generally overstated.

On the other hand, whereas children's concern about credibility appears to be driven largely by analytic credibility evaluation processes (which involve the effortful and deliberate
consideration of information), those who find Internet information most credible use more heuristic (hasty and feeling-based)
processes to evaluate it. This finding, coupled with the fact that
most kids said that people should be concerned about the credibility of information online, suggests that while kids take the
issue of credibility seriously, actual decisions about credibility
are not always based on a stringent approach to evaluating the
information they find online.

In addition, children report being equally likely to believe
entertainment and health information online, which implies
potentially problematic outcomes since these types of information should typically warrant different levels of skepticism. Also,
children consistently overestimate their own skill levels and
capacity to discern good from bad information as compared to
others. Such overconfidence is troubling, inasmuch as it implies
a correspondingly reduced level of vigilance or attention. And,
although most children displayed a healthy level of skepticism toward the hoax sites presented to them in this study,

approximately 10 percent of the children still believed the information on these hoax sites "a lot" or "a whole lot." Findings such as these illustrate that although youth exhibit encouraging signs of achieving appropriate skills and attitudes about online information credibility, there remain important gaps in their knowledge and abilities.

A number of factors appear to partially explain the coexistence of these encouraging and discouraging findings. For example, as kids get older, their Internet use increases both in scope and in time spent online. This increase may be due in part to decreased regulation by parents, and is accompanied by an increase in the variety of tools used to assess the credibility of information online. Older teens also trust the Internet more as an information source than do younger kids but think that people should be more concerned about the quality of information online than do younger children. This might indicate that as kids become more experienced with the Internet they have a greater appreciation for the potential of deceptive information online as well as greater confidence in their ability to find credible information sources.

Indeed, various forms of experience play a critical role in youth's credibility perceptions and information evaluation behaviors. Kids who have been using the Internet for a longer period of time, who spend a lot of time in virtual worlds, or who have contributed information to an online source (e.g., a blog, *Wikipedia*, etc.) think about credibility more and find more of the information and people they meet online to be credible. Also, older kids and kids who report having had or heard about bad experiences online report lower levels of belief in online information. In addition, our research indicates a positive relation between experience using the Internet and the use of analytic strategies for assessing the credibility of online information: as kids become more experienced using the Internet, they show more concern for the believability of information online, use more cognitively demanding tools to assess its credibility, and show a higher level of trust toward people and information online. When it comes to actively processing credibility cues to assess information credibility, practice appears to reap real rewards.

Implications and Future Directions

Findings from this study reveal a relationship between youth, the Internet, and credibility that is far more nuanced than previous research has suggested. Our study indicates that a combination of experience using the Internet over time and vital cohort-related changes in youth's cognitive development interact to promote better awareness of general credibility concerns and the ability to evaluate information found online. This has implications for several domains, including education and the creation of media literacy curricula, children's use of the Internet, policy formulations, and future research endeavors.

For example, based on our findings, online media literacy programs should emphasize a structured but graduated approach to guiding children's use of the Internet, which stresses the accumulation of personal experience online, early parental involvement, and the sharing of positive and negative online experiences at an early age. Curricula should be developed with these factors in mind, and should be assessed in terms of developmental and experiential differences among children.

This study also indicates that although overall experience may be a good predictor of credibility concern, it may also lull youth and even parents into believing they are better at discerning the credibility of information online than they actually are. Therefore, educational efforts regarding credibility evaluation should be ongoing, and should be targeted at youth with varying experience and skill levels in order to remain relevant. Indeed, quite different approaches appear warranted for younger versus older children and for those with lower versus higher online experience and skill.

It is also important to note that a number of limitations inherent in the survey methodology color our findings. For example, as use accumulates over time, children appear to appraise their ability to discern good versus bad information inaccurately. However, since survey data cannot accurately assess people's actual ability to find credible information successfully, techniques other than surveys should be used to validate and reveal any biases that result from this overconfidence. A possible direction for future research into this area is to investigate youth's evaluation of consequential information, such as information sought for schoolwork, during an actual information-seeking task. This could be done experimentally, via observation, or by other means.

As another example of the limitation of the survey method, we relied on screenshots of Web pages in our quasi-experiments to represent actual Web pages. Although this method has the advantage of experimental control, it suffers from its non-naturalistic nature. The screenshots simply cannot fully represent children's actual information seeking or browsing experiences, which would require methods that retain the context of such experiences. Considering this context could, of course, affect the results presented here, in ways that are not entirely predictable.

Additionally, while it appears most kids were appropriately concerned about the believability of the hoax Web sites represented in this study, there is a need to determine what characteristics and contextual factors led the minority to believe this information, above and beyond simple ignorance that may remain irrespective of the presentation of information via the Internet. To better understand this outcome, future research may investigate the effect of developmental states from age 11 forward, for example, on evaluating Web site credibility.

Overall, the findings presented here not only represent the current state of knowledge on this topic, but also serve as an important springboard for future research. Based on our findings, research should consider the development of children's information evaluation styles and strategies over time, differences in and the effects of parental involvement, the role of negative experiences online, the evolution and influence of false confidence in information evaluation abilities, and the most appropriate educational efforts to enhance and assess online information literacy.

Conclusion

One goal of this study was to move away from the simplistic treatments children often receive in examinations of youth and digital media, which cast children as either substantially more tech-savvy than adults, and therefore as superior in their use of digital media, or as universally vulnerable, and therefore in need of constant protection. Such accounts are prone to unnecessarily provoke either alienation or outrage, depending on the perspective taken. Our data suggest that neither view is particularly warranted, and that children's relation to digital media with regard to credibility is significantly more nuanced than either of these positions suggests.

In the end, and in spite of some evidence to the contrary, the reality seems to be largely what we would hope for as citizens, fellow Internet users, and parents: children are for the most part aware of the issues surrounding information verity on the Internet and appear generally capable of making informed and appropriate decisions in this regard. Thus, the best strategy to help children become more skillful Internet information consumers would appear to be from a perspective that empowers them and capitalizes on their unique upbringing in an all-digital world. Indeed, in a future in which the information that drives their lives is assembled, transmitted, shared, and processed digitally, children need to develop the skills necessary to navigate that information environment effectively. Perhaps the most encouraging conclusion from our data so far is that, for the most part, children seem to be making inroads toward that goal.

Appendixes

Appendix A: List of Tables and Figures

Tables

Figures

Appendix B: Knowledge Networks Methodology and Panel Recruitment

Knowledge Networks has recruited the first online research panel that is representative of the entire U.S. population. Panel members are randomly recruited by probability-based sampling, and households are provided with access to the Internet and hardware if needed.

Knowledge Networks selects households using random digit dialing (RDD) and address-based sampling methods. Once a person is recruited to the panel, they can be contacted by email (instead of by phone or mail). This permits surveys to be fielded very quickly and economically. In addition, this approach reduces the burden placed on respondents, since email notification is less obtrusive than telephone calls, and most respondents find answering Web questionnaires to be more interesting and engaging than being questioned by a telephone interviewer.

Beginning recruitment in 1999, Knowledge Networks established the first online research panel (now called Knowledge-Panel®) based on probability sampling that covers both the online and offline populations in the United States. The panel members are randomly recruited by telephone and by self-administered mail and Web surveys. Households are provided with access to the Internet and hardware if needed. Unlike other Internet research that covers only individuals with Internet access who volunteer for research, Knowledge Networks surveys are based on a dual sampling frame that includes both listed and unlisted phone numbers, telephone and non-telephone households, and cell-phone-only households. The panel is not limited to current Web users or computer owners. All potential panelists are randomly selected to join the KnowledgePanel; unselected volunteers are not able to join.

Random-Digit-Dialing Sample Frame

Knowledge Networks initially selects households using random-digit-dialing (RDD) sampling and address-based sampling (ABS) methodology. In this section, we will describe the RDD-based methodology, while the ABS methodology is described in a separate section below.

KnowledgePanel recruitment methodology uses the quality standards established by selected RDD surveys conducted for the federal government (such as the Centers for Disease Control-sponsored National Immunization Survey).

Knowledge Networks utilizes list-assisted RDD sampling techniques based on a sample frame of the U. S. residential landline telephone universe. For efficiency purposes, Knowledge Networks excludes only those banks of telephone numbers (a bank consists of 100 numbers) that have fewer than two directory-listings. Additionally, an oversample is conducted among a stratum of telephone exchanges that have high concentrations of

African American and Hispanic households based on census data. Note that recruitment sampling is done without replacement, thus numbers already fielded do not get fielded again.

A telephone number for which a valid postal address can be matched occurs in about 70 percent of the sample. These address-matched cases are all mailed an advance letter informing them that they have been selected to participate in KnowledgePanel. For efficiency purposes, the unmatched numbers are under-sampled at a current rate of 0.75 relative to the matched numbers. Both the over-sampling mentioned above and this under-sampling of non-address households are adjusted appropriately in the panel's weighting procedures.

Following the mailings, the telephone recruitment begins for all sampled phone numbers using trained interviewer/recruiters. Cases sent to telephone interviewers are dialed for up to 90 days, with at least 14 dial attempts on cases where no one answers the phone, and on numbers known to be associated with households. Extensive refusal conversion is also performed. The recruitment interview, about 10 minutes long, begins with informing the household member that they have been selected to join KnowledgePanel. If the household does not have a computer and access to the Internet, they are told that in return for completing a short survey weekly, they will be provided with a laptop computer (previously a WebTV device was provided) and free monthly Internet access. All members in a household are then enumerated, and some initial demographic and background information on prior computer and Internet use are collected.

Households that inform interviewers that they have a home computer and Internet access are asked to take their surveys using their own equipment and Internet connection. Per survey incentive points, redeemable for cash, are given to these "PC" respondents for completing their surveys. Panel members who were provided with either a WebTV or a laptop computer (both with free Internet access) do not participate in this per survey points incentive program. However all panel members do receive special incentive points for select surveys to improve response rates and for all longer surveys as a modest compensation of burden.

For those panel members receiving a laptop computer (as with the former WebTV), prior to shipment, each unit is custom configured with individual email accounts, so that it is ready for immediate use by the household. Most households are able to install the hardware without additional assistance, though Knowledge Networks maintains a telephone technical support line. The Knowledge Networks Call Center also contacts household members who do not respond to email and attempts to restore contact and cooperation. PC panel members provide their own email addresses, and weekly surveys are sent to that email account.

All new panel members receive an initial survey to both welcome them as new panel members and familiarize them with how online survey questionnaires work. They also complete a separate profile survey that collects essential demographic information such as gender, age, race, income, and education to create a personal member profile. This information can be used to determine eligibility for specific studies, is used for weighting purposes, and operationally need not be gathered with each and every survey. (This information is updated annually with each panel member.) Once new members are "profiled," they are designated as "active" and ready to be sampled for client studies. (*Note*: Parental or legal guardian consent is also collected for conducting surveys with teenage panel members, age 13–17.)

Once a household is contacted by phone—and additional household members recruited via their email address—panel members are sent surveys linked through a personalized email invitation (instead of by phone or mail). This permits surveys to be fielded quickly and economically, and also facilitates longitudinal research. In addition, this approach reduces the burden placed on respondents, since email notification is less obtrusive than telephone calls, and allows research subjects to participate in research when it is convenient for them.

Address-Based Sampling (ABS) Methodology

When Knowledge Networks started KnowledgePanel® panel recruitment in 1999, the state of the art in the industry was that probability-based sampling could be cost effectively carried out using a national random-digit-dial (RDD) sample frame. RDD at the time allowed access to 96 percent of the U.S. population. This is no longer the case. They introduced the ABS sample frame to rise to the well-chronicled changes in society and telephony in recent years that have reduced the long-term scientific viability of the RDD sampling methodology: declining respondent cooperation to telephone surveys; do not call lists; call screening, caller-ID devices, and answering machines, dilution of the RDD sample frames as measured by the working telephone number rate; and finally, the emergence of households that no longer can be sampled by RDD—the cell-phone-only households (CPOHH).

According to the Centers for Disease Control, approximately 21 percent of U.S. households cannot be contacted through RDD sampling: 18 percent as a result of CPOHH status and 3 percent because they have no phone service whatsoever. Among some segments of society, the sample non-coverage is substantial: almost one-third of young adults age 18–24 reside in CPOHHs. After conducting an extensive pilot project in 2008, Knowledge Networks made the decision to add an address-based sample (ABS) frame in response to the growing number of cell-phone-only households that are outside of the RDD frame. Before conducting the ABS pilot, they also experimented with supplementing their RDD samples with cell-phone samples. However, this approach was not cost effective and raised a number of other operational, data quality, and liability issues (e.g., calling people's cell phones while they were driving, for example).

The key advantage of the ABS sample frame is that it allows sampling of almost all U.S. households—an estimated 99 percent of U.S. households are "covered" in sampling nomenclature. Regardless of households' telephone status, they can be reached and contacted. Second, the ABS pilot project revealed some other advantages beyond the expected improvement in recruiting adults from CPOHHs as well:

• Improved sample representativeness for minority racial and ethnic groups.

• Improved inclusion of lower educated and low income households.

• Exclusive inclusion of CPOHHs that have neither a landline telephone nor Internet access (approximately 4 percent to 6 percent of U.S. households).

ABS involves probability-based sampling of addresses from the U.S. Postal Service's Delivery Sequence File. Randomly sampled addresses are invited to join KnowledgePanel through a series of mailings and in some cases telephone follow-up calls to non-responders when a telephone number can be matched to the sampled address. Invited households can join the panel by one of several means:

• Completing and mailing back a paper form in a postage-paid envelope.

• Calling a toll-free hotline maintained by Knowledge Networks.

• Going to a designated Knowledge Networks Web site and completing the recruitment form.

As mentioned earlier, after initially accepting the invitation to join the panel, respondents are then profiled online by answering demographic questions and maintained on the panel using the same procedures established for the RDD-recruited research subjects. Respondents not having an Internet connection are provided a laptop computer and free Internet service. Respondents sampled from the RDD and ABS frames are provided the same privacy terms and confidentiality protections that Knowledge Networks has developed over the years and have been reviewed by dozens of institutional review boards.

Because Knowledge Networks has recruited panelists from two different sample frames—RDD and ABS—they take several technical steps to merge samples sourced from these frames. This approach preserves the representative structure of the overall panel for the selection of individual client study samples. An advantage of mixing ABS frame panel members in any KnowledgePanel sample is a reduction in the variance of the weights. An ABS-sourced sample tends to align more true to the overall population demographic distributions, and thus the associated adjustment weights are somewhat more uniform and less varied. This variance reduction efficaciously attenuates the sample's design effect and confirms a real advantage for study samples drawn from KnowledgePanel with its dual frame construction.

Notes

Andrew Flanagin is a Professor in the Department of Communication at the University of California, Santa Barbara, where he also serves as the Director of the Center for Information Technology and Society. Dr. Flanagin's research focuses on the ways in which communication and information technologies structure and extend human interaction, with particular emphases on credibility, collective organizing, social media, and collaborative groups.

Miriam Metzger is an Associate Professor in the Department of Communication at the University of California, Santa Barbara. Dr. Metzger's research focuses on how information consumers assess the credibility of information in the new media environment and on issues of online information privacy.

Drs. Flanagin and Metzger are recognized as leading experts on credibility and digital media. They were among the first to conduct and publish empirical research on this topic, and have written several articles and reviews on Web credibility over the last decade, including the most comprehensive treatment of credibility in the online context to date. They have served as expert advisors on credibility for several organizations, including the MacArthur Foundation, the American Library Association, the Center for Media Literacy, and the National Library of Medicine. In addition, Drs. Metzger and Flanagin recently received a grant from the General Program of the MacArthur Foundation for their study "Credibility and Digital Media: Helping People Navigate Information in the Digital World" and coedited the volume *Digital Media, Youth, and Credibility* (MIT Press, 2008) as part of the MacArthur Foundation Series on Digital Media and Learning.

Ethan Hartsell, Alex Markov, Ryan Medders, Rebekah Pure, and Elisia Choi are graduate students in the Department of Communication, at the University of California, Santa Barbara.

1. Benchmark distributions for Internet access among the U.S. population of adults are obtained from KnowledgePanel recruitment data since this measurement is not collected as part of the Current Population Survey.

2. Since Knowledge Networks does not collect profile data for 11- and 12-year-olds, to set up the benchmarks of those with Internet access, they first weighted all 13- to 18-year-olds to look like the 11- to 18-year-old general population using Current Population Survey benchmarks. Thirteen-year-olds were treated as if they were 11 and 12 years old; thus 13-year-olds were weighted to be 36.17 percent of this population instead of 15.53 percent within all profiled members ages 13 to 18. Then, based on the weights for all 13- to 18-year-old KnowledgePanel members, Knowledge Networks derived the benchmarks based on those who have Internet access from home and weighted the child respondents to these Internet benchmarks.

3. The scales were constructed by relying on the results of principal components factor analyses, and were informed by factor loadings and the face validity of the questions we asked on the survey.

4. Multiple regression analysis was used to produce all results presented in this section. Detailed statistical information is available from the authors.

5. It is interesting that younger children said they were more likely to believe information they find online than did older children in light of our earlier finding that older children said they believe more of the information on the Internet than do younger children (see the section,

"Perceived Trust and Credibility of Web-based Information"). This could be due to younger children's realization that they are particularly susceptible to believing misinformation online, and older children's greater accumulation of positive experiences online, in terms of finding information that is useful and credible—a few of the factors that played into kids' tendency to trust (or not trust) people they encountered online.

6. The Web sites were modified slightly from their original online versions for size.

References

Alesina, A., and E. La Ferrara. 2002. "Who Trusts Others?" *Journal of Public Economics* 85: 207–234.

Alexa. 2009. "Top Sites: The Top 500 Sites on the Web." http://www.alexa.com/topsites (accessed on August 25, 2009).

Alexander, J. E., and M. A. Tate. 1999. *Web Wisdom: How to Evaluate and Create Information Quality on the Web.* Hillsdale, NJ: Erlbaum.

Benkler, Y. 2006. *The Wealth of Networks: How Social Production Transforms Markets and Freedom.* New Haven: Yale University Press.

Campbell, J., N. Greenauer, K. Macaluso, and C. End. 2007. "Unrealistic Optimism in Internet Events." *Computers in Human Behavior* 23: 1273–1284.

Chesney, T. 2006. "An Empirical Examination of Wikipedia's Credibility." *First Monday* 11, no. 11. http://outreach.lib.uic.edu/www/issues/issue11_11/chesney/index.html (accessed on October 30, 2009).

Clarke, V. A., H. Lovegrove, A. Williams, and M. Macpherson. 2000. "Unrealistic Optimism and the Health Belief Model." *Journal of Behavioral Medicine* 23, no. 4: 367–376.

Eastin, M. 2008. "Toward a Cognitive Developmental Approach to Youth Perceptions of Credibility." In *Digital Media, Youth, and Credibil-*

ity, ed. M. J. Metzger and A. J. Flanagin, 29–47. Cambridge, MA: MIT Press.

Epstein, S., R. Pacini, V. Denes-Raj, and H. Heier. 1996. "Individual Differences in Intuitive-Experiential and Analytical-Rational Thinking Styles." *Journal of Personality and Social Psychology* 71, no. 3: 390–405.

Flanagin, A. J., C. Flanagin, and J. Flanagin. 2010. "Technical Code and the Social Construction of the Internet." *New Media & Society* 12, no. 2: 179–196.

Flanagin, A. J., and M. J. Metzger. 2000. "Perceptions of Internet Information Credibility." *Journalism & Mass Communication Quarterly* 77: 515–540.

Flanagin, A. J., and M. J. Metzger. 2007. "The Role of Site Features, User Attributes, and Information Verification Behaviors on the Perceived Credibility of Web-based Information." *New Media & Society* 9, no. 2: 319–342.

Flanagin, A. J., and M. J. Metzger. 2008. "Digital Media and Youth: Unparalleled Opportunity and Unprecedented Responsibility." In *Digital Media, Youth, and Credibility*, ed. M. J. Metzger and A. J. Flanagin, 5–27. Cambridge, MA: MIT Press.

Giles, J. 2005. "Internet Encyclopaedias Go Head to Head." *Nature* 428 (December 15): 900–901.

Jacobs, J. E., and P. A. Klaczynski, eds. 2005. *The Development of Judgment and Decision Making in Children and Adolescents*. Mahwah, NJ: Erlbaum.

Jenkins, H. 2006. *Convergence Culture: Where Old and New Media Collide.* New York: New York University Press.

Kamins, M. A., M. J. Brand, S. A. Hoeke, and J. C. Moe. 1989. "Two-Sided versus One-Sided Celebrity Endorsements: The Impact on Advertising Effectiveness and Credibility." *Journal of Advertising* 18, no. 2: 4–10.

Klaczynski, P. A. 2001. "The Influence of Analytic and Heuristic Processing on Adolescent Reasoning and Decision Making." *Child Development* 72: 844–861.

Kokis, J., R. Macpherson, M. Toplak, R. F. West, and K. E. Stanovich. 2002. "Heuristic and Analytic Processing: Age Trends and Associations with Cognitive Ability and Cognitive Styles." *Journal of Experimental Child Psychology* 83: 26–52.

Krane, B. 2006 (November 13). "Researchers Find Kids Need Better Online Academic Skills." *The Advance.* http://advance.uconn.edu/ 2006/061113/06111308.htm (accessed on August 6, 2009).

Lenhart, A., and M. Madden. 2007 (April). *Teens, Privacy, and Online Social Networks.* Pew Internet & American Life report. http:// www .pewinternet.org/Reports/2007/Teens-Privacy-and-Online-Social-Net-works. aspx?r=1 (accessed on August 25, 2009).

Lenhart, A., Madden, M., & Hitlin, P. 2005 (July). *Teens and Technology: Youth Are Leading the Transition to a Fully Wired and Mobile Nation.* Pew Internet & American Life report. http://www.pewinternet.org/report _display.asp?r=162 (accessed on April 17, 2006).

Leu, D. J., J. Coiro, J. Castek, D. Hartman, L. A. Henry, and D. Reinking. 2008. "Research on Instruction and Assessment in the New Literacies of Online Reading Comprehension." In *Comprehension Instruction: Research-Based Best Practices,* ed. C. Collins Block, S. Parris, and P. Afflerbach, 321–346. New York: Guilford.

Metzger, M. J. 2007. "Making Sense of Credibility on the Web: Models for Evaluating Online Information and Recommendations for Future Research." *Journal of the American Society for Information Science and Technology* 58, no. 13: 2078–2091.

Metzger, M. J., and A. J. Flanagin, eds. 2008. *Digital Media, Youth, and Credibility.* Cambridge, MA: MIT Press.

Metzger, M. J., A. J. Flanagin, K. Eyal, D. R. Lemus, and R. McCann. 2003. "Credibility in the 21st Century: Integrating Perspectives on

Source, Message, and Media Credibility in the Contemporary Media Environment." In *Communication Yearbook 27*, ed. P. Kalbfleisch, 293–335. Mahwah, NJ: Erlbaum.

Metzger, M. J., A. J. Flanagin, and R. B. Medders. Forthcoming. "Social and Heuristic Approaches to Credibility Evaluation Online." *Journal of Communication*.

Metzger, M. J., A. J. Flanagin, and L. Zwarun. 2003. "College Student Web Use, Perceptions of Information Credibility, and Verification Behavior." *Computers & Education* 41, no. 3: 271–290.

Quantcast. 2009. "Quantcast US Site Rankings." http://www.quantcast.com/top-sites-1 (accessed on August 26, 2009).

Rainie, L. 2006. "Life Online: Teens and Technology and the World to Come." Keynote address, annual conference of the Public Library Association, Boston, MA, March 23. http://www.pewinternet.org/ppt/Teens%20and%20technology.pdf (accessed on November 7, 2006).

Rieh, S. Y., and B. Hilligoss. 2007. "College Students' Credibility Judgments in the Information Seeking Process." In *Digital Media, Youth, and Credibility*, ed. M. J. Metzger and A. J. Flanagin, 49–71. Cambridge, MA: MIT Press.

Scott, S. G., and R. A. Bruce. 1995. "Decision-Making Style: The Development and Assessment of a New Measure." *Educational and Psychological Measurement* 55, no. 5: 818–831.

Slater, M. D., and D. Rouner. 1997. "How Message Evaluation and Source Attributes May Influence Credibility Assessment and Belief Change." *Journalism & Mass Communication Quarterly* 73: 974–991.

Tapscott, D. 1997. *Growing Up Digital: The Rise of the Net Generation*. New York: McGraw-Hill.

van Dijk, J. 2006. "Digital Divide Research, Achievements and Shortcomings." *Poetics* 34: 221–235.

Weinstein, N. D. 1980. "Unrealistic Optimism about Future Life Events." *Journal of Personality and Social Psychology* 39, no. 5: 806–820.

Weinstein, N. D. 1982. "Unrealistic Optimism about Susceptibility to Health Problems." *Journal of Behavioral Medicine* 5, no. 4: 441–460.

Weinstein, N. D. 1987. "Unrealistic Optimism about Susceptibility to Health Problems: Conclusions from a Community-Wide Sample." *Journal of Behavioral Medicine* 10, no. 5: 481–500.